Berichte aus dem
Institut für Umformtechnik
der Universität Stuttgart
Herausgeber: Prof. Dr.-Ing. K. Lange

74

Grundlagen der Umformtechnik – Stand und Entwicklungstrends

Fundamentals of Metal Forming Technique – State and Trends

Vorträge des Internationalen Symposiums
Stuttgart, 13./14. Oktober 1983

Proceedings of the International Symposium
Stuttgart, October 13/14, 1983

Teil/Part I

Springer-Verlag
Berlin Heidelberg New York Tokyo 1983

Dr.-Ing. Kurt Lange

o. Professor an der Universität Stuttgart
Institut für Umformtechnik

ISBN-13: 978-3-540-13039-0   e-ISBN-13: 978-3-642-82186-8
DOI: 10.1007/978-3-642-82186-8

Telefon 0 70 33/38 25-26
2362/3020—543210

Vorwort / Preface

Die Referate des Symposiums befaßten sich mit den Themenkrei-
sen Plastizitätstheorie, Tribologie, Werkstofftechnik, Tech-
nologie des Umformens, rechnerunterstützte Prozeßsimulation
sowie mit CAD/CAM für Werkzeuge und Maschinen. Die meisten
Beiträge stammen von anerkannten Wissenschaftlern aus der
Bundesrepublik Deutschland, Dänemark, Frankreich, Großbritan-
nien, Japan, Kanada, Österreich, Schweden,der Schweiz und den
USA. In weiteren Kurzbeiträgen wurde über abgeschlossene und
laufende Arbeiten des Institutes für Umformtechnik der Universität Stuttgart berichtet.

Alle Vorträge sind in der Originalsprache, d. h. entweder in
Deutsch oder in Englisch abgedruckt.

Die Veranstaltung wandte sich an das Fachmanagement und an
Ingenieure in Forschung und Entwicklung, Konstruktion und Ar-
beitsvorbereitung in Unternehmen, die sich mit der Umformtechnik befassen sowie an Mitarbeiter von Hochschulinstituten
und Mitglieder technisch-wissenschaftlicher Vereinigungen.

The papers read at the symposium covered the topics theory of
plasticity, tribology, materials, technology of metal forming,
computer aided process simulation as well as CAD/CAM for tools
and machine tools. In addition findings of completed or current investigations carried out at the Institut für Umformtechnik der Universität Stuttgart were presented.

All papers are printed in their original language only, either
English or German.

The conference was addressed to technical managers and to
professional engineers in research and development, design
and process planning in industrial companies dealing with
metal forming as well as to scientific university staff and
members of scientific technical societies.

<u>Inhaltsverzeichnis / Table of Contents</u>   Seite / Page

Autorenliste / List of Authors   9

The Cylindrical Bending of Strips and the Biaxial
Stamping of Rectangular Plates into Curved Dies
W. Johnson, T. X. Yu   11

Finite Element Process Modeling of Sheet Metal
Forming of General Shapes
C. H. Toh, S. Kobayashi   39

A Review of Development and Use of Upper Bound
Approach to Metal Forming Processes
H. Kudo   57

Entwicklung von Stoffgesetzen für die Hochtemperaturplastizität
E. Steck   83

The Industrial Use of Computer Progams in Cold
Forging
J. F. Renaudin, H. J. Braudel   115

Some Key Problems in the Tribology of Metal-
forming
J. A. Schey   139

Optimale Reibbedingungen beim geschmierten Tiefziehprozess
J. Reissner   151

Simulation with Model Materials in the Nordic
Countries
T. Wanheim   167

Einfluß von Wechseln der Beanspruchungsrichtung
auf die Fließspannung von Metallen
H. P. Stüwe   189

Die Bedeutung der Umformverfahren für die systematische Werkstoffauswahl
P. Huml   201

Seite / Page

Über die Wechselwirkung von Werkstoff und Umformung und ihre Beschreibung durch Versuche
K. Pöhlandt   215

Turbinenscheiben für Flugtriebwerke: Beanspuchung,
Werkstoffe, Fertigungsverfahren
H. Wilhelm   271

Autorenliste / List of Authors

Prof. W. Johnson, em. Professor der Mechanik, University of Cambridge, Great Britain

Dr. T. X. Yu, Department of Engeneering, University of Cambridge, Great Britain

Dr. C. H. Toh, Department of Mechanical Engeneering, University of California, Berkeley, USA

Prof. S. Kobayashi, Department of Mechanical Engeneering, University of California, Berkeley, USA

Prof. Dr.-Eng. H. Kudo, Department of Mechanical Engeneering, Yokohama National University, Japan

Prof. Dr.-Ing. E. Steck, Institut für Allgemeine Mechanik und Festigkeitslehre (Mechanik B), Technische Universität Braunschweig, Bundesrepublik Deutschland

Dr. J. F. Renaudin, C.E.T.I.M., Saint Etienne, Frankreich

H. J. Braudel, C.E.M.E.F., Sofia Antipolis, Frankreich

Prof. J. A. Schey, Department of Mechanical Engeneering, University of Waterloo, Kanada

Prof. Dr. J. Reissner, Institut für Umformtechnik, Eidgenössische Technische Hochschule Zürich, Schweiz

Prof. T. Wanheim, Afdelningen for Mekanisk Teknologi, Danmarks Tekniske Hojskole, Lyngby, Dänemark

Prof. Dr. H. P. Stüwe, Erich-Schmid-Institut für Festkörperphysik, Österreichische Akademie der Wissenschaften, Leoben, Österreich

Prof. P. Huml, Working of Metals, Royal Institute of Technology, Königlich Technische Hochschule Stockholm, Schweden

Dr.-Ing. K. Pöhlandt, Institut für Umformtechnik, Universität Stuttgart, Bundesrepublik Deutschland

Dr.-Ing. H. Wilhelm, Motoren- und Turbinen-Union, München, Bundesrepublik Deutschland

Dr.-Ing. G. Schröder, BBC Aktiengesellschaft, Baden, Schweiz

Dr.-Ing. K. Roll, Control Data GmbH, Stuttgart, Bundesrepublik Deutschland

Dipl.-Ing. M. Dostal, Institut für Umformtechnik, Universität Stuttgart, Bundesrepublik Deutschland

Dipl.-Ing. H. Noller, Institut für Umformtechnik, Universität Stuttgart, Bundesrepublik Deutschland

Prof. Dr.-Ing. M. Geiger, Lehrstuhl für Fertigungstechnologie, Friedrich-Alexander-Universität Erlangen-Nürnberg, Bundesrepublik Deutschland

Dipl.-Ing. W. König, Lehrstuhl für Fertigungstechnologie, Friedrich-Alexander-Universität Erlangen-Nürnberg, Bundesrepublik Deutschland

Dipl.-Ing. E. Dannenmann, Institut für Umformtechnik, Universität Stuttgart, Bundesrepublik Deutschland

Dr.-Ing. R. Geiger, Press- und Stanzwerk AG, Eschen, Liechtenstein

Dipl.-Ing. W. Schätzle, Gebr.Felss, Königsbach-Stein, Bundesrepublik Deutschland

Prof. Dr.-Ing. D. Schmoeckel, Institut für Umformtechnik, Technische Hochschule Darmstadt, Bundesrepublik Deutschland

Prof. Dr.-Ing. F. Dohmann, Universität-Gesamthochschule-Paderborn, Bundesrepublik Deutschland

Dr.-Ing. T. Neitzert, Daimler-Benz AG, Stuttgart, Bundesrepublik Deutschland

Dr.-Ing. H. Glöckl, Institut für Umformtechnik, Universität Stuttgart, Bundesrepublik Deutschland

Prof. Dr. T. Altan, Battelle Columbus Labs., Ohio, USA

Dr. S. I. Oh, Battelle Columbus Laboratories, Ohio, USA

The Cylindrical Bending of Strips and the Biaxial Stamping of Rectangular Plates into Curved Dies

W. Johnson and T. X. Yu,
Engineering Department, Cambridge University, Cambridge, U.K.

Summary

Strips of copper, brass and mild steel, horizontally and freely
supported in a circular die, were formed using a matching punch.
The developments of clearance between the strip and the punch
pole, and of primary and secondary bottoming of the strips
against the die with consequent multi-point bending are described. The associated punch load-travel curves are shown and
compared with the results of a theoretical analysis. The
effects of tensile forces and strain-hardening on bending and
springback are also summarized.

Representative results are presented concerning the stamping of
rectangular plates into doubly-curved dies by matching punches
in order to examine the effects of die curvature ratio, specimen
dimension and the largest punch load applied. Photographs are
shown of the contact regions which develop between plate, die
and punch. Rigid/plastic deformation models with appropriate
hinge lines are proposed and the experimental results are seen
to be in good accord throughout.

List of Symbols

a_1, a_2	half side-lengths of rectangular plate
D	diameter
E	Young's modulus
E_p	strain-hardening modulus
h	thickness
L	half length of strip
M_e	largest elastic bending moment
M_p	fully plastic bending moment
P	punch load
$P*$	largest punch load applied
R	radius of curvature
S	area of plate
s	arc length
Y	yield stress
Δ	punch travel
η	non-dimensional parameter, μ_m/μ_b
κ	curvature
λ	non-dimensional parameter, L/R
μ_b, μ_m	non-dimensional parameters
ξ	standard deviation of curvature distribution
ρ	non-dimensional parameter, YR/Eh

ω curvature-ratio of die, κ_1^D/κ_2^D

Introduction

The bending of strips, sections and plates into singly-curved and doubly-curved dies respectively by matching dies (in the case of the first and last mentioned workpieces) is a subject of very great industrial interest. A major practical concern in such bending processes is with the springback or elastic recovery which occurs in a product when the forming load is removed. Product dimensions are effectively and usually pre-determined by designers and there is thus a need to be able to calculate amounts of springback in order to initially specify forming die and punch magnitudes. This need is however one which researchers have devoted little attention to satisfying. Whilst aiming mainly to investigate the latter topic however it is not the only one to which we have addressed ourselves, and indeed in the Summary above the many facets of both kinds of pressing as encountered and investigated, are listed. The latter are only shortly reported in respect of both theoretical and experimental work but references to complete detailed accounts are, or soon will be, available elsewhere. It is the authors' hope that sufficient is disclosed to enable readers to appreciate the whole gamut of problems encountered with both types of bending situation.

Cylindrical Bending of Strips

Most previous researchers on the bending and springback of strips and sheets have examined V-die bending or flange-forming. However, in industrial practice metal strips or sheets are also often brought to a cylindrical shape by stamping with a cylindrical punch into a cylindrical die. Previous work on some facets of this has been described by Queener and de Angelis, Dannenmann, Weinmann and Shippell, Sidebottom and Gebhardt, see ref. [1].

Johnson and Singh [2] presented the first comprehensive research (known to the authors) on the cylindrical bending of metal strips. They reported experimental work in which strips were formed into a cylindrical die by a cylindrical punch to study the dependence of springback (the average final curvature) on the tool radius, the dimensions of the strips and the properties of materials.

Yu and Johnson [3] carried on this research and found new interesting features in the cylindrical bending process. Figure 1 shows the experimental setup they used. It consists of a gauge-holder fixed to the cross-head of an Instron testing machine and a dial gauge which was able to travel down with the punch during the test, the reading on the dial gauge indicating the clearance between the punch pole and the mid-point of a specimen.

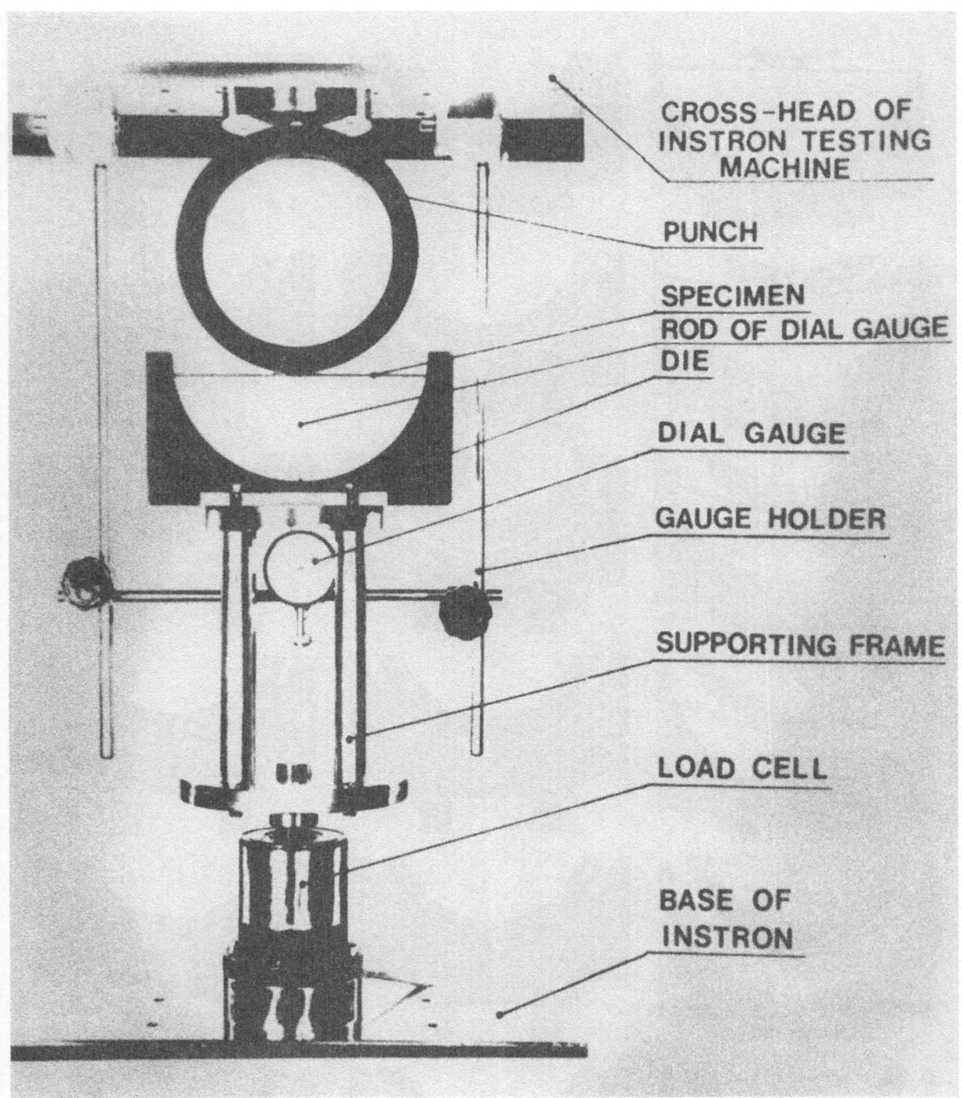

Fig. 1: The experimental setup

Copper, brass and mild steel specimens were used in the experiments. A typical sequence in the deformation process is as shown in Fig. 2, with consecutive loading/bending configurations, Nos. 1-14, and the unloading/springback process, Nos. 15-18, for a copper specimen of length 9.9 in (251.5 mm) and thickness 1/4 in (6.35 mm), deformed between a die (D_{die} = 10 in = 254 mm) and a punch (D_{punch} = D_{die} - 2h = 241.3 mm). In Fig. 2,

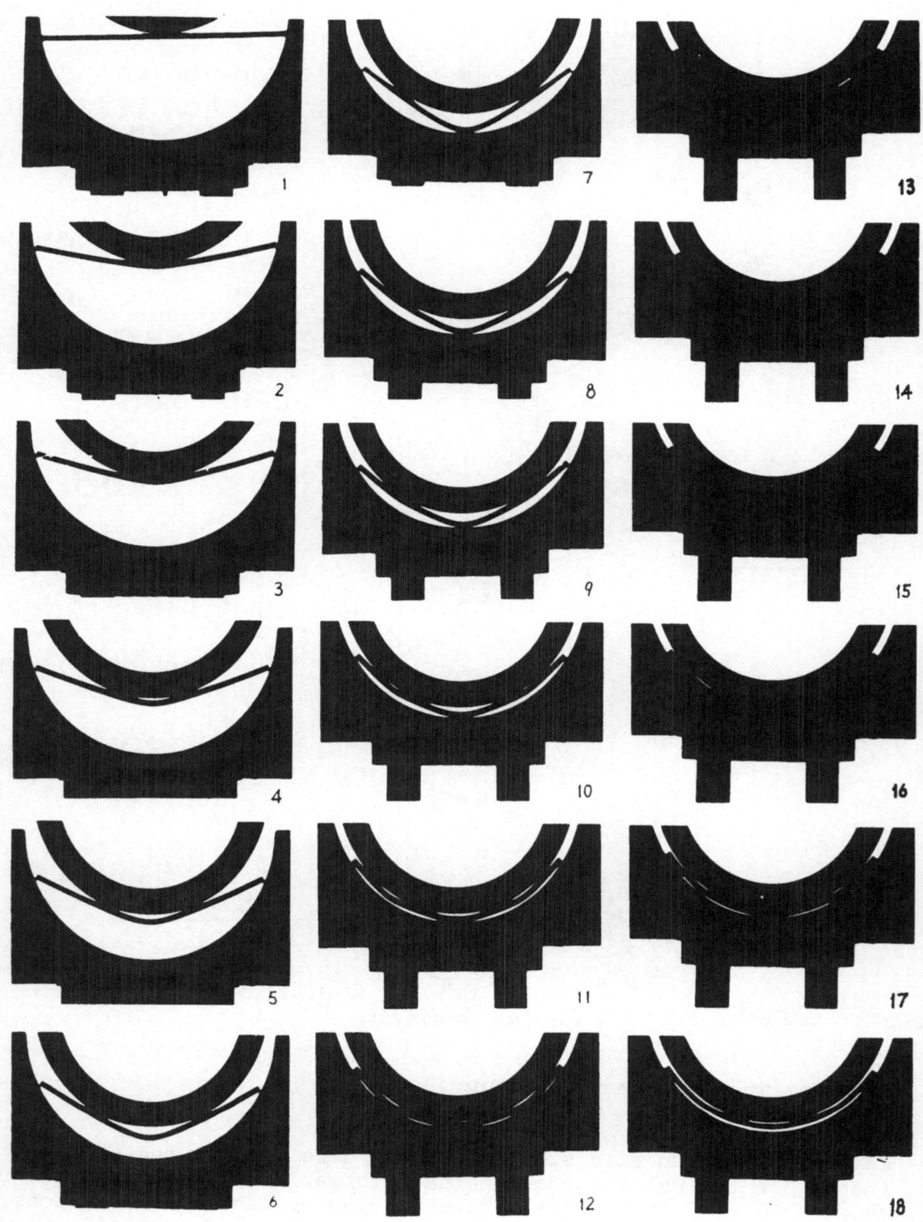

Fig. 2: The deformation process for a copper specimen
D_{die} = 254 mm, 2L = 251.5 mm, h = 6.35 mm

No. 1 shows the initial state.

No. 2 shows bending of the strip, without separation from the punch pole; this is 3-point bending.

Nos. 3 to 6 show the onset and the increase in separation between the punch pole and the middle part of the strip as bending progresses; this is 4-point bending.

No. 7 reveals primary bottoming of the mid-point of the strip on the die; this is 5-point bending.

Nos. 8 and 9 show more bending on the two sides of the strip - still 5-point bending.

Nos. 10 and 11 show the onset and development of secondary separation between the punch and the strip on the two sides; this is now 7-point bending.

No. 12 reveals the secondary bottoming of the strip on to the die - an increase to 9-point bending.

Nos. 13 to 15 show the gaps between the punch and the strip reducing as well as that between the die and the strip, as the strip is pressed further - a multi-point bending situation.

Nos. 16 and 17 show the unloading process with fewer and fewer contact regions as the punch is gradually withdrawn. (It will be observed that there is in fact some degree of asymmetry in the disposition of the points of contact.)

No. 18 shows the final state when springback is completed.

Experimental results in total indicate that, (i) among the strips made of the same material, the thinner strips display a higher degree of elastic behaviour during bending and undergo a larger amount of springback after unloading; and (ii) in comparison with brass strips, copper and mild steel strips undergo a larger separation from the punch during bending and less springback after unloading.

Typical punch load-travel curves obtained in the tests are shown in Fig. 3, where P denotes the punch load and Δ the punch travel. In Stage I of the process, the load increases with punch travel at a nearly linear rate because the strip is in an elastic state. After the onset of plasticity, i.e. in Stage II, the load-carrying capacity of the strip is nearly constant over a long range. In Stage III, i.e. after the mid-point of the strip bottoms on the die, the punch load undergoes a sharp increase. The details of the $P \sim \Delta$ characteristics after Stage III can be seen in Fig. 3(b), which was obtained by adopting a very low cross-head speed.

The clearance between the punch pole and the mid-point of the strip depends on the dimensions and material of strips. Fig.

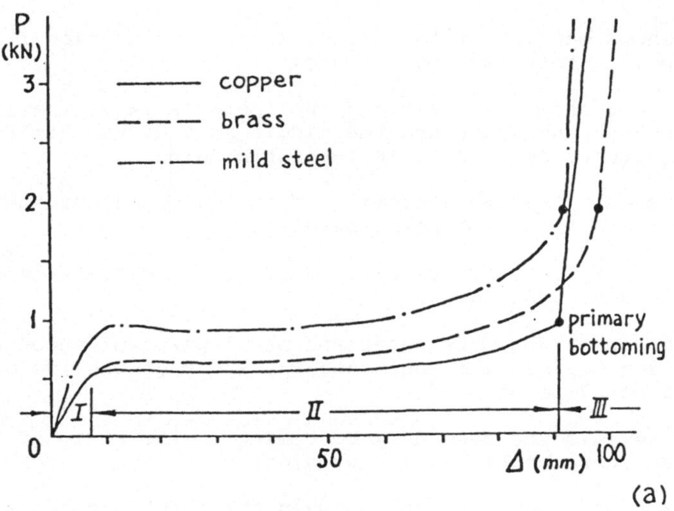

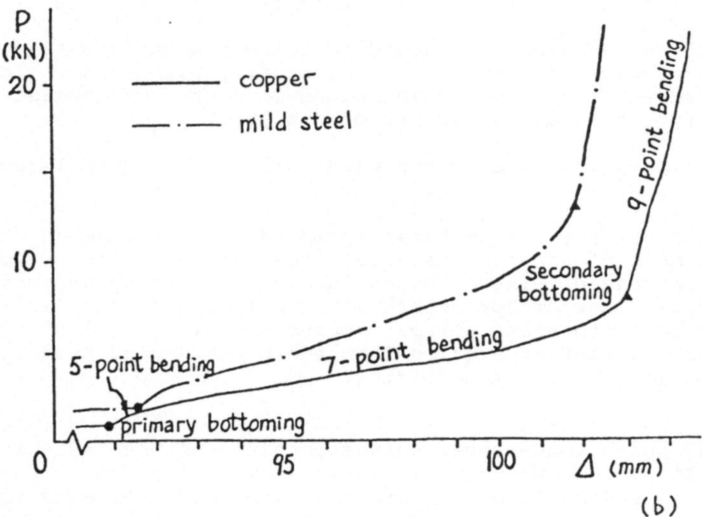

Fig. 3: The punch load - travel curves
D_{die} = 254 mm, 2L = 251.5 mm, h = 6.35 mm
(a) the whole curve;
(b) detailed behaviour in the later stage.

4(a) shows that among strips of the same size, the clearance
for copper is larger than that for mild steel, whilst the
clearance for brass is much smaller than either. Figure 4(b)
indicates that among strips made of the same material (copper,

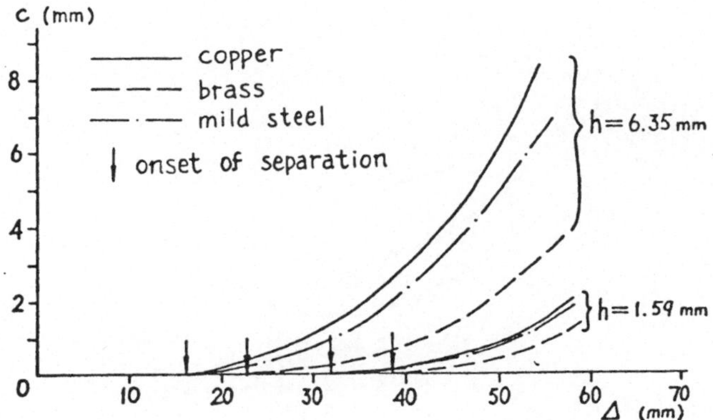

Fig. 4(a): The clearances between the punch pole
and the mid-point of strips
D_{die} = 152.4 mm, 2L = 149.9 mm

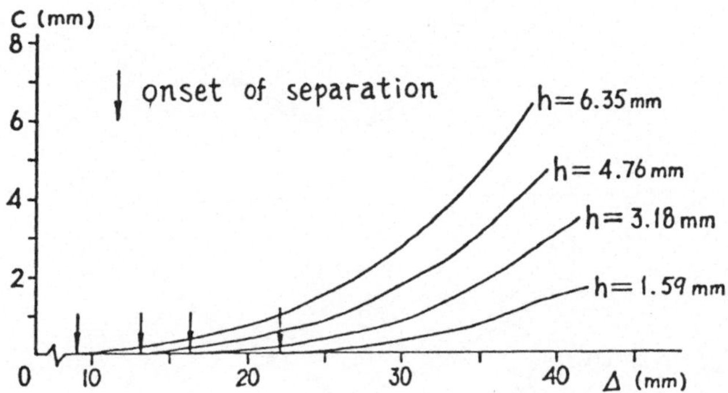

Fig. 4(b): The clearances between the punch pole
and the mid-point of the strips of copper.
D_{die} = 114.3 mm, 2L = 111.8 mm

in this example) and having the same length, the thicker strip
has a larger clearance c, and it separates from the punch earlier.

After loading to a certain largest punch-load p* and completely
unloading, the strip specimens assume their final shapes with
more or less springback, and some examples are shown in Fig. 5.
Figs. 5(a) and 5(b) show the effects of the relative thickness
and of the material on springback, respectively, whilst Fig.
5(c) provides evidence of the importance of the largest punch
load applied on springback. Indeed, for a relatively small
largest punch-load, the strip has some parts that are nearly
straight. Conversely, if the largest punch-load is great
enough, then the final shape of the strip is approximately

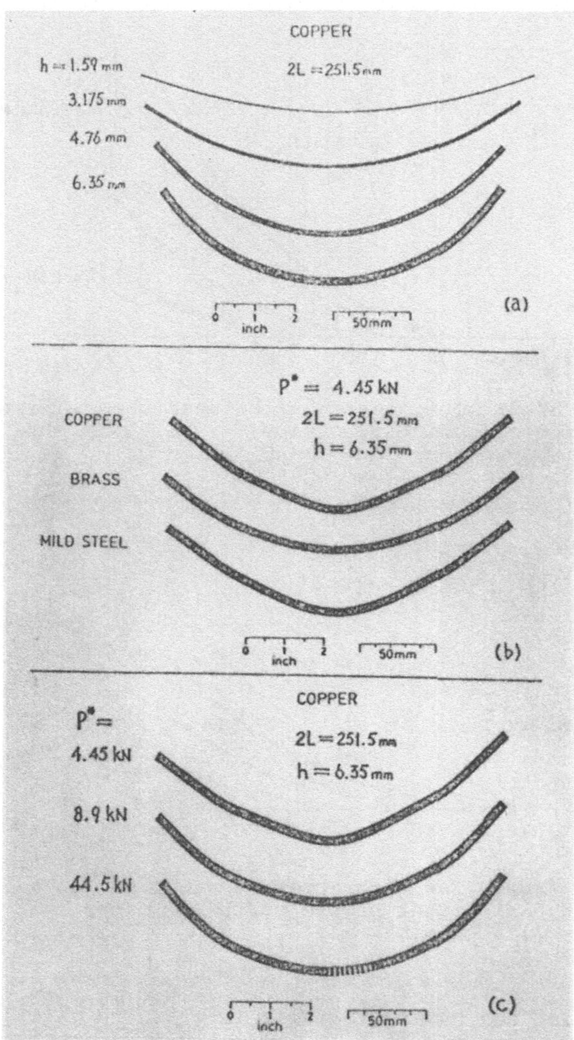

Fig. 5: Photographs of specimens after unloading
(a) copper specimens of different thicknesses;
(b) specimens of different materials;
(c) copper specimens loaded to different largest
punch-loads

circular, although the radius may be different from those of
the punch and die.

After measuring the profiles of the bent strips, the distributions of the curvatures along the arc length were calculated by using a numerical procedure. In Fig. 6, κ denotes the final curvature of strips and s denotes the arc length measured from the mid-point towards one of the ends. Notably, the curvature distribution depends very much on the largest punch load applied, as shown in Fig. 6(b).

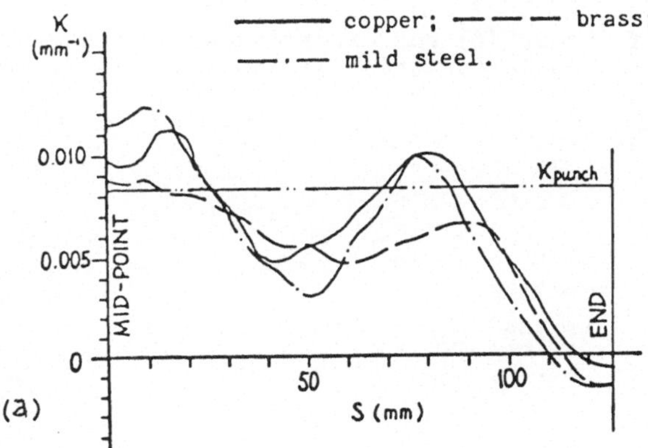

Fig. 6(a): The distribution of curvature κ, D_{die} = 254 mm, 2L = 251.5 mm, h = 6.35 mm.
p* = 8.9 kN, s denotes the arc-length measured from the mid-point of the strip

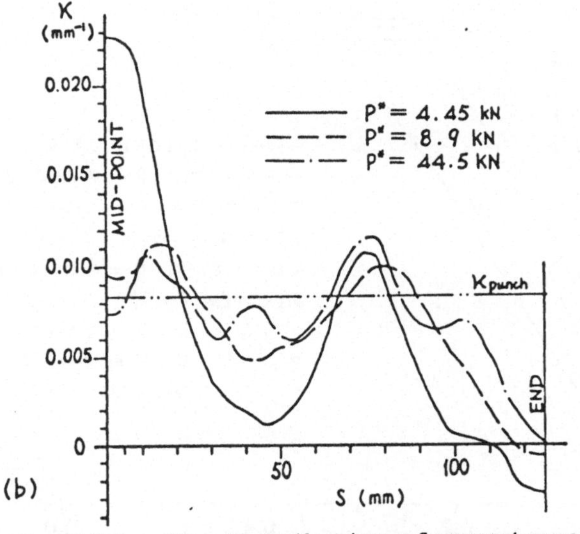

Fig. 6(b): The distribution of curvatures of copper specimens, 2L = 251.5 mm, h = 6.35 mm.

The corresponding theoretical analysis [4] is carried out stage-by-stage for the deformation process. For instance, during 3-point and 4-point bending one half of the strip can be regarded as an elastic-plastic cantilever loaded by an inclined force at its tip. Also, a rigid/plastic analysis can be performed by assuming one plastic hinge (for 5-point bending) or more plastic hinges (for 7- or multi-point bending).

A comparison of theory [4] and experiments [3] concerning the punch load-travel characteristics is shown in Fig.7, where

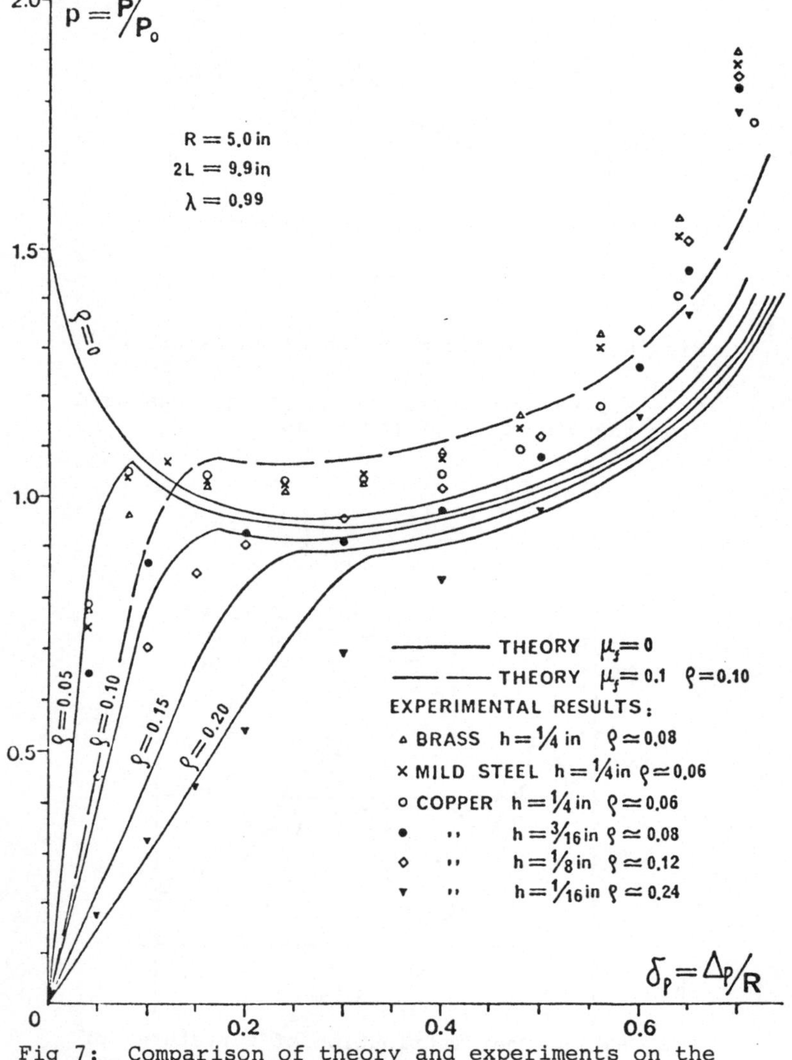

Fig 7: Comparison of theory and experiments on the punch load-travel characteristics

$P_0 = 2M_e/L$ and $M_e = Yh^2/6$; $2L$ and h are the length and thickness
of the strip respectively and R is the average radius of the
punch and die. It has been confirmed by our theoretical analysis
that two non-dimensional parameters, i.e. $\rho = (Y/E)(R/h)$ and
$\lambda = L/R$, dominate the deformation modes and the stage divisions;
the case $\rho = 0$ describes a rigid/perfectly plastic strip. μ_f
is the friction coefficient between the strip and the die.

Figure 8 is a comparison of theory [4] and experiments [3] on
the clearance between the punch pole and the mid-point of strips.
The experimental values are less than those theoretically
predicted, but this can be explained away by reference to the
strain-hardening properties of the materials used.

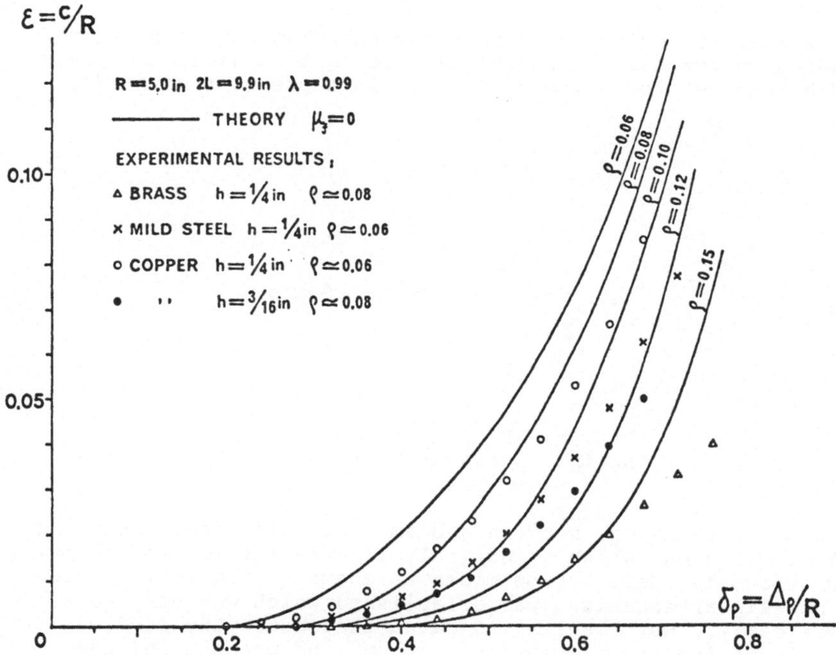

Fig. 8: Comparison of theory and experiments on the
clearance between the punch pole and the mid-point
of strips.

From both experiments and theoretical analysis it is concluded
that the deformation process and springback of strips in cylindrical bending are dominated by (YR/Eh) and (L/R) as well as the
largest punch load applied. It also emerges that in order to
produce a nearly perfect final shape, the largest punch load
applied should be 20 to 30 times that required to enforce the
onset of plasticity.

In [5], we analysed the large elastic deflection of strips in
V-die bending according to the theory of the Elastica. It has
been shown that neglect of the elastic deflection in thin strips
may lead to a notably incorrect prediction of its load-carrying
capacity. From this situation the condition for the onset of
plasticity in strips has also been set up.

When a strip is subjected to a bending moment as well as an
axial force, its neutral axis will be displaced from the position it adopted when subjected to only a bending moment. In
a previous work [6] an extensive study was carried out into the
influence of axial force on the elastic-plastic bending of a
strip and on the resulting springback of the strip. The results
indicate that the effect of an axial force is to decrease the
amount of springback.

To examine the effect of strain-hardening in sheet or strip-
bending, a typical and useful example is to analyse the press-
brake bending process, see Fig. 9. In a previous work [7],

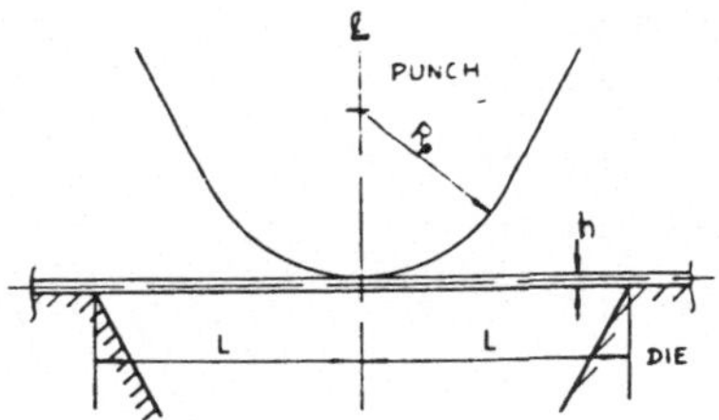

Fig. 9: Punch, die and undeformed
strip: press-brake forming

by using the analogy between a beam of rigid/linear hardening
material and one which is entirely linear-elastic, and adopting
two successive deformation modes as shown in Fig. 10, a system
of non-linear equations was established which was easy to solve
numerically. Our results indicated that after the middle
portion of the strip is wrapped around the punch pole, the
curvature of the strip at the mid-point becomes larger than
the curvature of the punch, whilst that at the contact point,
A, becomes smaller than the punch curvature. Accompanying
this variation in curvature, the middle portion of the strip
separates from the punch pole, and the clearance depends on
the dimension of the strip as well as the strain-hardening
rate of the material. Figure 11 shows profiles of the bent
strip with $L/R = 1.5$ and $(Y/E_p)(R^2/Lh) = 1$, where Y is yield
stress and E_p the strain-hardening modulus. This figure also
indicates the movement of contact point A and of point B, the
boundary between the plastic and rigid regions during the
forming process.

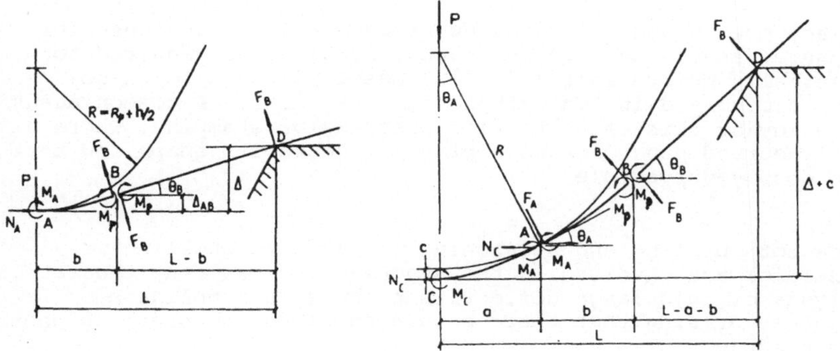

Fig. 10(a): Deformation mode I   Fig. 10(b): Deformation mode II

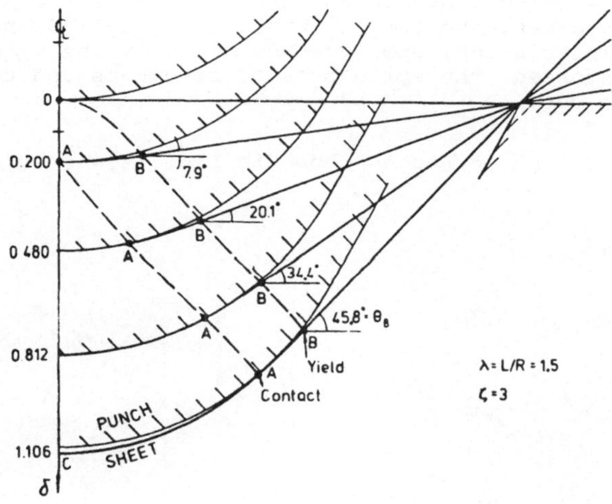

Fig. 11: The profile for bent strip

Our analysis indicates that for the middle portion of the strip
in the press-brake bending, the ratio of average curvatures
after and before springback can be estimated as

$$\frac{\overset{F}{K}}{K} = 1 - \left(\frac{E_p}{E} + 3\,\frac{YR}{Eh} \right).$$

Thus, the higher the yield stress and the higher the strain-
hardening modulus the more the springback.

On comparing our results with those obtained by the finite
element method, see Oh and Kobayashi's work [8], the degree of
agreement is found to be very good.

The Stamping of Rectangular Plates into Doubly-Curved Dies

In many present-day fabricating processes, e.g. in those for
pressure vessels and storage tanks, it is often required to
plastically deform initially flat metal plates into shapes
which are curved in two orthogonal directions. A common means
of achieving this is by biaxial pressing or stamping, where a
doubly-curved punch forces a plate to match its shape and that
of a doubly-curved die.

Since both bending and stretching (as well as buckling or
wrinkling) are involved in the stamping process, the theoretical
analysis of this large deflection process is complicated. It
is not surprising that there should be almost no previous papers
on the subject.

Recently, a series of experiments on the double-curvature
stamping of rectangular metal plates has been conducted [9] to
investigate the effects of the curvature-ratio of the die, the
size of the specimen, the largest punch load applied and the
material characteristics, upon the mechanism of the large
deformation enforced, the springback of specimens and the
wrinkling developed.

The experimental set-up was as shown in Fig. 12,whilst the

(a)

(b)

Fig. 12: The
experimental set-up
(a) without specimen;
(b) with a specimen.

1. the cross-head of the
    Instron testing machine
2. doubly-curved punch
3. plate specimen
4. doubly-curved die
5. load cell
6. the base of the Instron

shape of the doubly-curved die is shown in Fig. 13. The specimens were of mild steel and were cut from rolled sheets of 1.6 mm thickness.

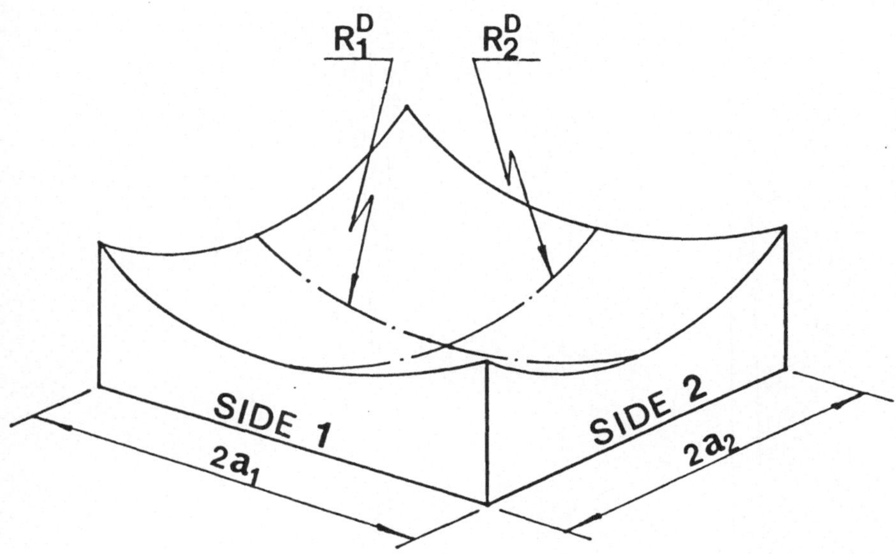

Fig. 13: The dimensions of the doubly-curved die.

(1) The punch load - travel curves

The experimental non-dimensional punch load-punch travel curves depend on the dimensions of the specimens, see Fig. 14, as well as the curvature-ratio of the die, i.e. $\omega = \kappa_1^D/\kappa_2^D$, see Fig. 15.

(The punch load is non-dimensionalized by SY, S being the area of the plate and Y the plate yield stress.) It can be seen from Fig. 14 that for narrow plates, the punch load undergoes continuous increase with punch travel but for plates whose shape is nearly square the curves display a notable rise and fall between primary bottoming occurring on one side and secondary bottoming on the other.

(2) The contact regions

To show the development of contact regions, before stamping a specimen, its two surfaces were painted white, and two sheets of carbon paper were put to cover the surfaces. During stamping, the pressure between the tool and the specimen made black marks on both surfaces of the specimen. Figure 16 shows the contact regions between a plate of 200 x 150 x 1.6 mm and the tools where $\omega = 1/1$.

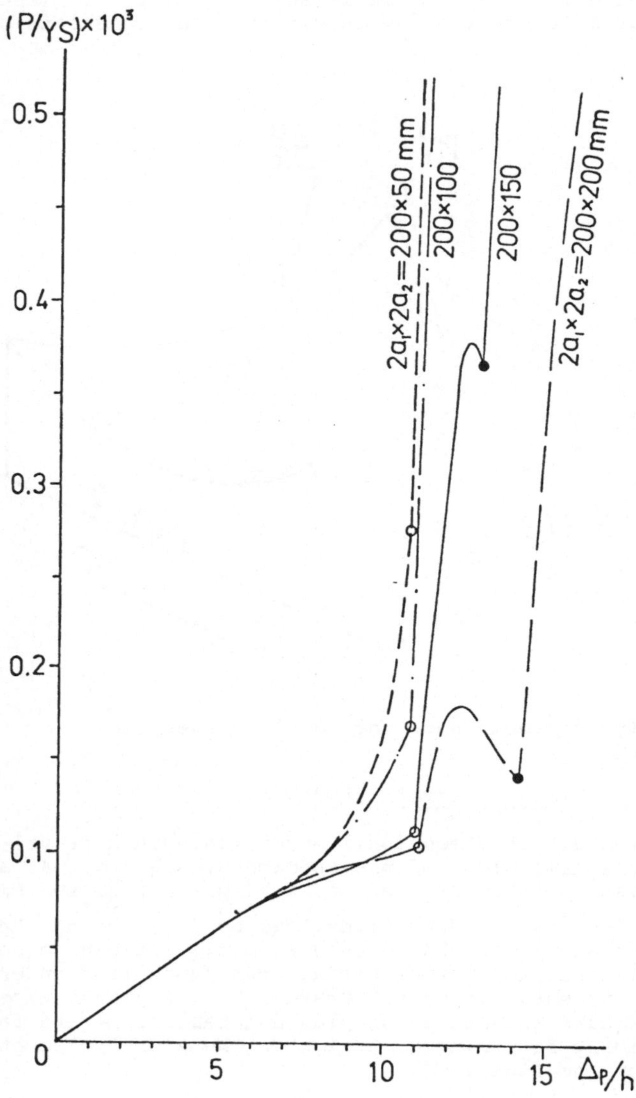

Fig. 14: The punch load - travel curves for the plates of various sizes pressed into the equi-curved die (i.e. $\omega = 1/1$)
o primary bottoming at mid-points of Side 1
• secondary bottoming at mid-points of Side 2.

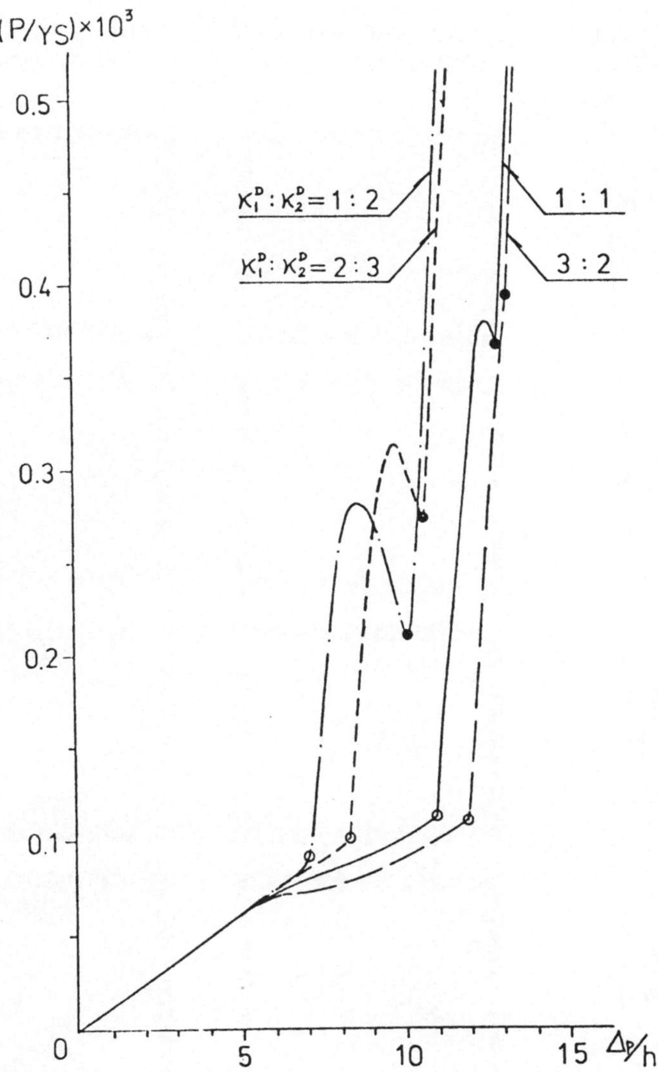

Fig. 15: The punch-load travel curves for
plates of 200 x 150 x 1.6 mm pressed into
different doubly-curved dies
o   primary bottoming at mid-points of Side 1
●   secondary bottoming at mid-points of Side 2

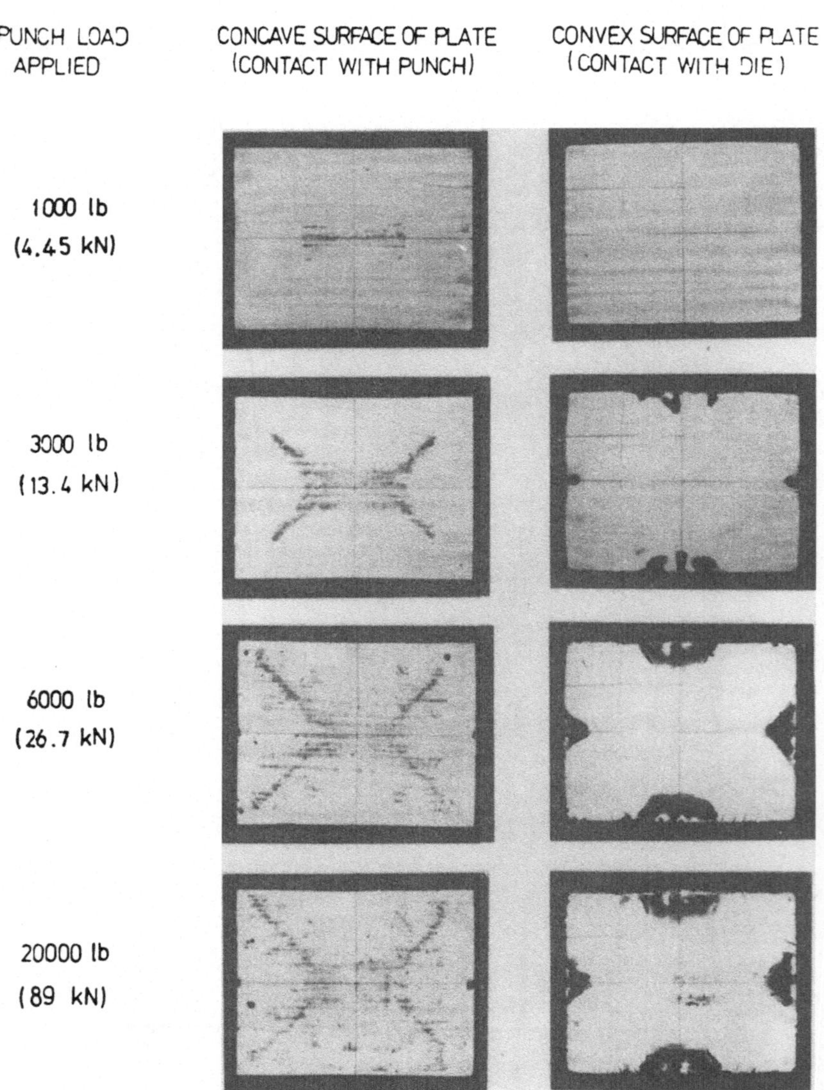

Fig. 16: Photographs showing the contact regions between a plate of 200 x 150 x 1.6 mm and the tools of $\omega = 1/1$

The development of the contact regions can be divided into three phases thus,

(i)   When the punch load was relatively small, say $P^*/YS \leq$ $\leq 0.5 \times 10^3$, the region in contact with the punch mainly extended somewhat along the longer centre line of the plate out from the centre; the regions in contact with the die were concentrated at the mid-points of the four edges of the plate.

(ii)   As the punch load increased, the region in contact with the punch changed to develop along the diagonals of the plate, or more appropriately, along the angle-bisectors of the four inner angles of the plate, providing a pattern similar to that which is associated with the plastic torsion sand-hill analogy for a shaft of this particular cross-sectional shape. The regions in contact with the die expanded in the neighbourhoods of the mid-points of the plate edges and appeared as wrinkles in those areas.

(iii)   When the punch load increased further, say $P^*/YS \geq 3 \times 10^{-3}$, the contact regions showed more branches on the concave surface and more small areas along the edges of the convex surface. This was particularly noticeable in a 'large' plate, where many wrinkles occurred along the edges.

(3)   The final curvature of plates

After stamping, the profiles of plate specimens were measured by traversing with a Ferranti SURFCOM along several lines parallel to the edges of each specimen. From these data, the distribution of the final curvature and therefore the average curvature and the corresponding standard deviations can be calculated. Figure 17 shows the results for a mild steel

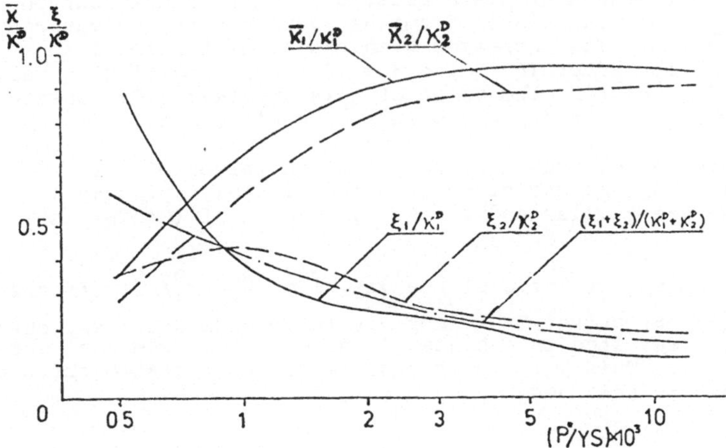

Fig.17:   The average final curvatures and their standard deviations for a plate of 200 x 150 x 1.6 mm pressed into the die of $\omega = 2/3$

specimen of 200 x 150 x 1.6 mm pressed into a die where ω = 2/3 and where $\bar{\kappa}_1$ and $\bar{\kappa}_2$ denote average values of the principal curvatures of the plate, and ξ_1 and ξ_2 are the corresponding standard deviations. This figure confirms that when the largest punch load P* increases while the area of the plate S is a constant, the final curvatures of the plate approach those of the die whilst the former become more uniformly distributed throughout the whole plate.

To find one parameter which will act as an index of the springback of plates, we write

$$\mu_b \equiv (\bar{\kappa}_1 + \bar{\kappa}_2)/(\kappa_1^D + \kappa_2^D),$$

$$\mu_m \equiv (\bar{\kappa}_1 \cdot \bar{\kappa}_2)/(\kappa_1^D \cdot \kappa_2^D)$$

and define

$$\eta \equiv \mu_m/\mu_b.$$

For an equi-biaxial stamping, i.e. where $\omega = \kappa_1^D/\kappa_2^D = 1/1$, $\eta = \bar{\kappa}/\kappa^D$ can be regarded as the 'springback ratio'. In general, for double - curvature stamping, η can be regarded as an index of the springback of a plate.

By means of the use of η, some experimental results can be compared. The overall effect of the non-dimensional largest punch load, P*/YS, on the springback of plates is shown in Fig. 18. From that Figure the following observations may be made:

(i) For a rectangular plate pressed into any doubly-curved die, the ratio η calculated from the average final curvatures of the pressed plate increases when the largest punch load increases; but after (P*/YS) $\simeq$ 3 x 10^{-3} (this particular value is for 1.6 mm plates) the value of η is approximately constant for a particular plate pressed into a specific die.

(ii) There is not much difference in the values of η for rectangular, square and circular plates if their principal dimensions are similar and if the same value of (P*/YS) is taken.

(iii) The curvature ratio of the die, i.e. $\omega = \kappa_1^D/\kappa_2^D$, affects the $\eta \sim$ (P*/YS) curve. It is not yet fully understood why our die for ω = 1/2 resulted in a higher value of η than the equi-curved die, ω = 1/1. This particular experiment was repeated but the latter feature remained.

(4) <u>Wrinkling phenomena</u>

In our experiments, some large specimens displayed wrinkling along their edges, as indicated in Fig. 16. The wrinkling waves

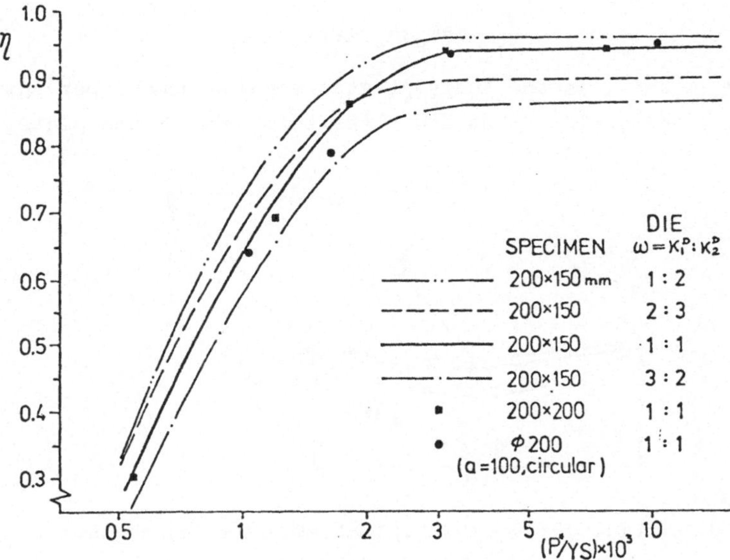

Fig. 18: The springback index η, affected by the non-dimensional largest punch load (P*/YS)

were approximately sinusoidal along the edges of plates. For the smaller specimens, no clear wrinkle was observed. The larger is the plate and the closer to square is the shape of plate, the more severe is the wrinkling.

To explain these experimental results, a series of approximate deformation models were developed [9] using the rigid/perfectly plastic material. These models made it possible to predict several features in the different stages of the stamping process, for example, the form of the punch load-travel characteristic, the change in contact regions and the development of wrinkling phenomena.

First, consider an incipient collapse mechanism. Since the punch first contacts the plate at its centre and the plate is initially supported at its four corners, the first phase of deformation is that of a rectangular plate of side-lengths $2a_1$ and $2a_2$, simply supported at the four corners and subjected to a concentrated force P at the centre. Due to the symmetry of the problem, only the case of $a_1 \geq a_2$ is considered below.

After examining several possible collapse mechanisms, it has been sufficiently substantiated that a mechanism with a unique hinge along the shorter central line, see Fig. 19, provides the lowest initial collapse force, i.e.

$$P_0 = 4M_p(a_2/a_1),$$

where $M_p = Yh^2/4$ is the fully plastic bending moment per unit length, Y the yield stress and h the thickness of the plate.

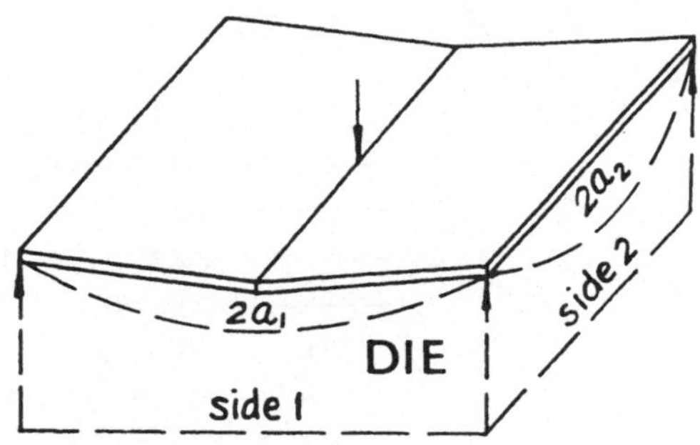

Fig. 19: The incipient collapse mechanism

In Stage I of the stamping process, i.e. from the beginning until primary bottoming occurs at the mid-points of the longer sides (side 1), bending mainly develops in side 1 whilst side 2 remains almost straight, that is, the plate undergoes approximately cylindrical bending during this stage. The elastic deformation in Stage I can be easily estimated by considering the plate to behave as a strip.

After primary bottoming, the plate is supported at six points, and there are then several possible succeeding deformation models. If the plate is narrow ($a_1/a_2 \geq 2$ or so), the cylindrical bending model may continue. In addition to the previously formed hinge along a central line, more hinges are assumed to form parallel to and symmetrically with respect to the central line, as shown in Fig. 20. Then, the analysis of this model is essentially the same as that we performed for the cylindrical bending of strips.

Another possible deformation model for Stage II (i.e. the stage in which the plate is supported at six points) is a four-segment deformation model, as shown in Fig. 21. However, one important difference between the four-segment model and the cylindrical bending model is that, as deformation proceeds, no membrane stress develops in the latter, whereas membrane stress unavoidably develops in the former. This results in additional load-carrying capacity for the four-segment model. We have proved for this model that [9],

$$P/P_0 \simeq 2 \cdot \frac{a_1}{a_2} \cdot \frac{\bar{\Delta}_p}{h},$$

where $\bar{\Delta}_p$ denotes the punch-travel at primary bottoming.

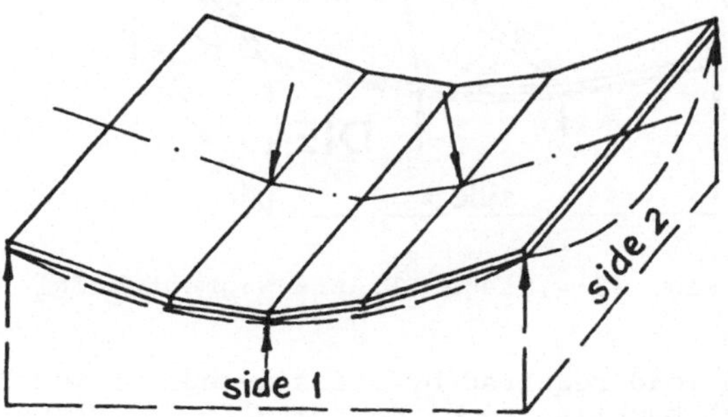

Fig. 20: Stage II, cylindrical bending model with three hinges

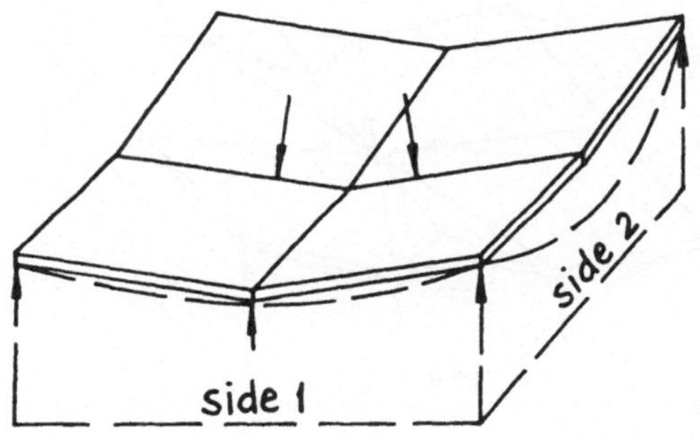

Fig. 21: Stage II, four-segment model

Therefore, if the aspect ratio of the plate, a_1/a_2, is close to unity, i.e. the plate is close to being square, then the four-segment model may require a lower load than the cylindrical model.

Our experimental observations strongly suggest that a six-segment model, soon succeeds the four-segment model. As shown in Fig. 22, this model has flattened triangular regions near the shorter central line. The inclined moving hinges cause these regions to expand without causing other portions of the plate to suffer membrane deformation. By estimating the energy dissipated plastically in the moving and fixed hinges, an increasing punch load-travel characteristic can be obtained [9].

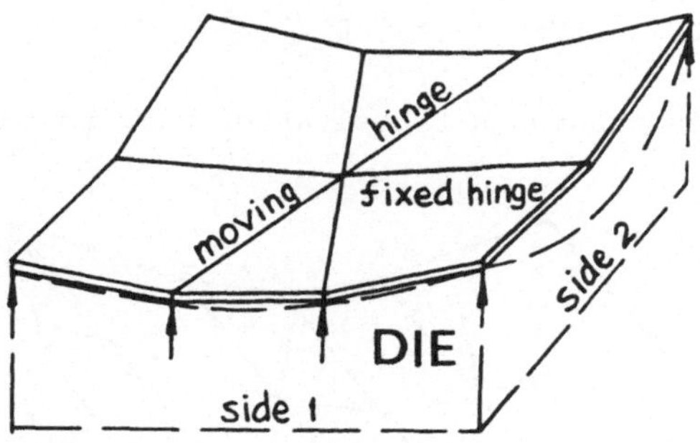

Fig. 22: Stage II, six-segment model

When the punch load required by the six-segment model is large
enough, a local buckling model may develop, as shown in Fig.
23; progress towards this model was the product of experimental
observation. After some mathematical treatment, a decreasing
$P \sim \Delta_p$ curve can be provided by this model [9].

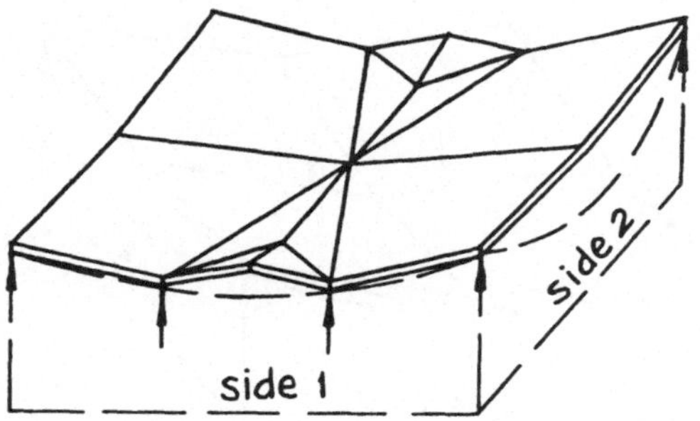

Fig. 23: Stage III, local buckling model

After secondary bottoming has occurred at the mid-points, B, of
the short sides (side 2), the plate is supported at 10 points.
As shown in Fig. 24, any further movement of region OBCE is
restricted and increments in punch-travel can be achieved only
if this region undergoes a shear membrane deformation. In this
stage a shear deformation model is required and the load-
carrying capacity may increase rapidly due to the action of
membrane forces.

To examine the deformation models proposed above, a numerical
calculation was carried out for a mild steel rectangular plate
of 300 x 250 x 1.6 mm pressed into the die of ω = 2/3 (R_1^D = 360
mm and R_2^D = 240 mm). As shown in Fig. 25, the agreement between

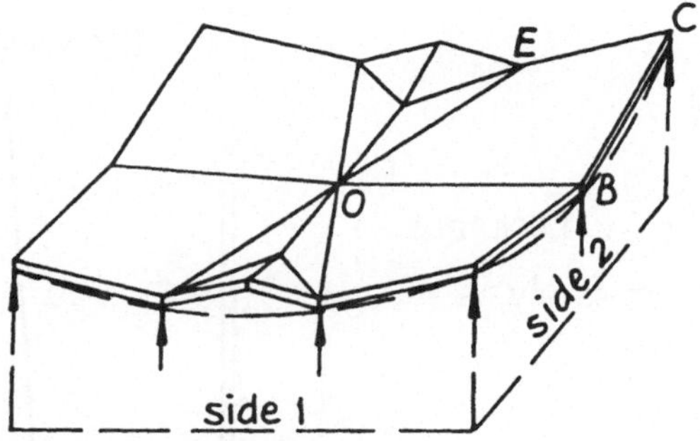

Fig. 24:   Stage IV, shear deformation model

the calculated results and the experimental curve was reasonably
good. The authors hope they will be forgiven for observing that
it is remarkable that the model analysis can follow the actual
variations in the load-carrying capacity so well in every stage
of the deformation process.

Conclusion

The theory of plastic bending forming processes tends to be one
which is regarded as largely worked out. We hope we have shown
however that this is not so and indeed we trust that students
of metal forming will not only find the results outlined to be
of significance for industrial and constructional application
but will also introduce some of the ideas into the teaching and
canon of standard, classical plasticity theory.

There are many directions in which the work reported above can
be usefully extended - e.g. to the cold and hot bending of
thick plates, the effects of deliberately introducing in-plane
forces, section-bending, composite plate bending, the direct
determination of residual stress distribution and to the
employment of other forms of both dies and shapes of strips and
plates. We hope that some of these will be pursued. Finally,
we trust that our paper will have done something towards
stimulating research in, and the re-examination of, an old and
historical forming process.

Acknowledgments

The authors are indebted to Dr. W. J. Stronge for his discussion
on deformation models of plates. They are also grateful to
Mrs S. Purlan for typing their manuscript.

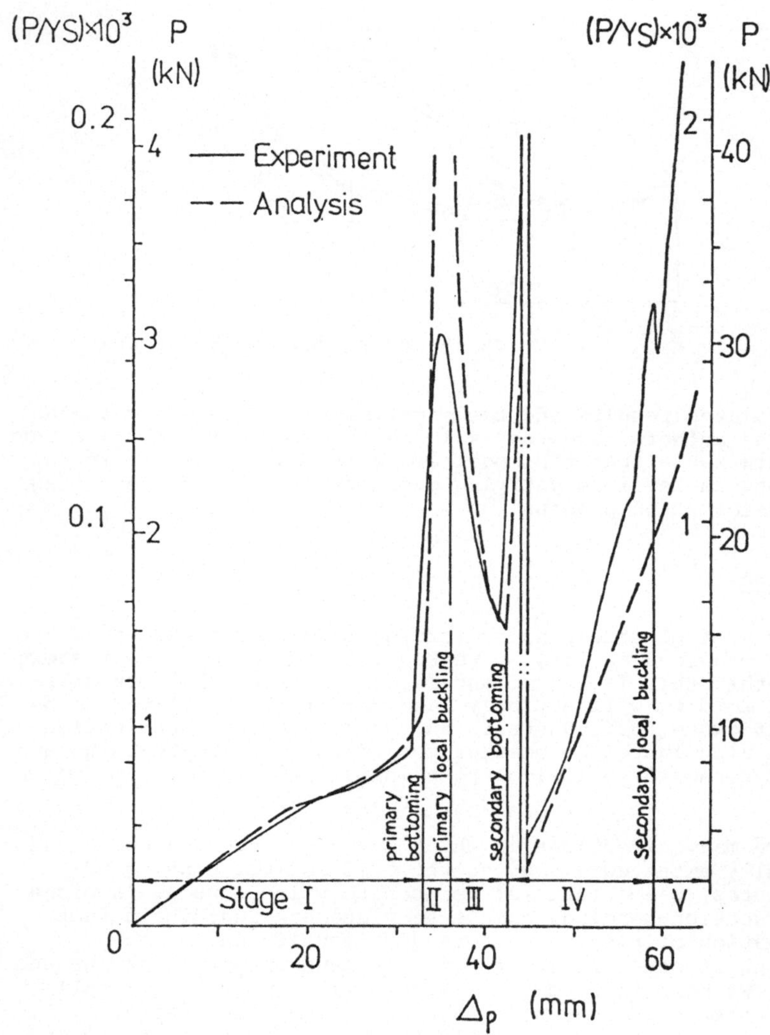

Fig. 25: Comparison of results of model analysis and experiment on the punch load - travel curve for a mild steel plate of 300 x 250 x 1.6 mm pressed into the die of ω = 2/3

References

[1] Johnson, W.: The mechanics of some industrial pressing, rolling and forging processes, in "Mechanics of Solids: The Rodney Hill 60th Anniversary Volume" (Eds. H. G. Hopkins

and M. J. Sewell), Pergamon Press (1982), pp. 303 - 356.

[2] Johnson, W.; Singh, A. N.: Springback after cylindrically
     bending metal strips, Conference on Large Deformation,
     Delhi, December, 1979, see 'Large Deformations', Edited by
     N. K. Gupta and S. Sengupta, South Asian Publishers (1982),
     pp. 236 - 250.

[3] Yu, T. X.; Johnson,W.: Experiments on the cylindrical
     bending of metal strips, to be published in Metals
     Technology.

[4] Yu, T. X.; Johnson, W.: A theoretical analysis on the
     cylindrical bending of strips, to be published in Metals
     Technology.

[5] Johnson, W.; Yu, T.X.: The large elastic deflection on
     thin strips in V-die bending and the onset of plasticity,
     to be published in Proceedings of Institution of Mechanical
     Engineers.

[6] Yu, T. X.; Johnson, W.: Influence of axial force on the
     elastic-plastic bending and springback of a beam, Journal
     of Mechanical Working Technology, (1982), $\underline{6}$, 5 - 21.

[7] Yu, T. X.; Johnson, W.: The press-brake bending of rigid/
     linear work-hardening plates, Int. J. Mech. Sci., (1981),
     $\underline{23}$, 307 - 318.

[8] Oh, S. I.; Kobayashi, S.: Finite element analysis of plane-
     strain sheet bending, Int. J. Mech. Sci., (1980), $\underline{22}$, 583
     - 594.

[9] Yu, T. X.; Johnson, W.; Stronge, W. J.: The stamping of
     rectangular plates into doubly-curved dies, to be published
     in the Int. J. Mech. Sci.

Finite Element Process Modeling of
Sheet Metal Forming of General Shapes

C. H. Toh, S. Kobayashi, Department of Mechanical Engeneering
University of California, Berkeley/U.S.A.

Summary

Sheet metal forming processes, axially symmetric and non-symmetric, were
analyzed numerically by the finite element method based on the membrane
theory. The finite element model takes into account the rigid-plastic
material characteristics which satisfy Hill's anisotropic yield criterion
and its associated flow rule. The theoretical formulation includes the
normal anisotropy of the sheet metal as well as the finite deformation
during the sheet forming process. In addition, Coulomb model was incorporated into the analysis as the interfacial friction condition between the
forming tool and the sheet metal. The punch stretching of a circular blank
with a hemispherical punch was first analyzed to examine the validity of
the developed finite element code. It was found that the computed solutions were consistent with the existing axisymmetric punch stretching
results. Then the solutions were obtained for non-symmetric sheet metal
forming problem of punch stretching of a rectangular strip with a hemispherical punch.

1   Introduction

In order to produce a useful sheet metal part, the initially simple geometry of sheet metals has to be transformed into a more complicated geometry
without material failure or deterioration in material properties. The process of sheet metal forming may involve, in general, local or general
bending, drawing, stretching, ironing or various combinations of these
modes of deformation. For sheet metals subjected to the same forming forces, the modes of deformation may differ appreciably due to the differences
in the material characteristics and the material-process interactions.
Successful forming, depends upon several factors, such as the formability
of the sheet metal itself, design of the part and the tooling, surface conditions of the sheet metal, lubrication, and the speed of the forming
press. Thus, the knowledge of the deformation mechanics and the mutual
interactions between the process parameters are of paramount importance in

controlling the metal flow efficiently during the operation and in providing a better design criterion for the actual sheet metal forming process. The efforts have already been carried out in analyzing various metal working processes, including sheet metal forming, using the finite element method.

Although quite a substantial amount of work on the numerical analysis of sheet metal forming processes has been done by many researchers in the past decade [1- 10], much of the effort was devoted to the axisymmetric problems with the infinitesimal strain formulation. The solution of a non-symmetric sheet forming problem has been obtained only recently by Onate and Zienkiewicz [11]. They derived the general shell element from the formulation based on the viscoplastic flow theory, and obtained the solution for the problem of hemispherical punch stretching of a square blank.

This paper describes a general finite element method to obtain detailed mechanics of deformation of rigid-plastic materials involved in various sheet metal forming processes with axisymmetric and non-symmetric sheet geometries. The finite element formulation takes into account the finite deformation with the updated geometry at the end of each incremental step. Coulomb friction is implemented as the tractional boundary condition and the membrane state of stress is assumed to exist within the domain of the sheet to be analyzed. The formulation was tested by comparing the axisymmetric solution for the problems of punch stretching and the new solution was obtained for stretching of a rectangular strip.

2 Kinematics of Finite Deformation and Virtual Work Principle

The material is assumed to be rigid-plastic and the formulation is based on the membrane theory. In order to establish the kinematical relationships between the successive stages of deformation of the sheet metal, a material point P_0, is considered at the undeformed configuration at time $t=0$ in the sheet metal which occupies a plane domain D contained in the X-Y plane of a fixed Cartesian coordinate system, X-Y-Z, as shown in Figure 1. It is supposed that, at a certain stage of deformation at time $t=t_0$, the material point, P_0, becomes P in a coordinate system x-y-z (reference configuration) and occupies the position $\underset{\sim}{X}_0$ with respect to the X-Y-Z system. Further imposing the displacement vector, $\underset{\sim}{u}$, the material point P will move again to another stage as P' at time $t=t_0 +\Delta t$, in a coordinate system x'-y'-z' (deformed configuration) and possess the position $\underset{\sim}{X}$ with respect

to the undeformed configuration. For the incremental deformation analysis, the position $\underset{\sim}{X}_0$ of the particle is assumed to be known or has been
calculated at the end of the previous increment. The problem to be solved
is to find the additional displacement vector, $\underset{\sim}{u}$, which satisfies the
equilibrium equation in the deformed configuration and the displacement
prescribed boundary conditions.

Because of the consideration of incremental deformation in the simulation
of the sheet metal forming processes, the primary attention is then focused
on the relationships between the stages at $t=t_0$ and at $t=t_0 + \Delta t$ with the
displacement vector, $\underset{\sim}{u}$, to be determined for each of the increment. In the
following formulation, the rectangular coordinate systems are used. The
formulations for general coordinate systems are given in Ref. [12]. Referring to Fig. 1, when an actual differential force vector, $\underset{\sim}{dF}$, is acting on
the area element, ds, of the material configuration at time $t=t_0 + \Delta t$,
there are various ways to denote this force. The actual force, $\underset{\sim}{dF}$, when
referred to the deformed configuration at $t=t_0 + \Delta t$, can be expressed as

$$d\underset{\sim}{F} = \sigma_{\alpha\beta}\, n_\alpha\, \underset{\sim}{g}_\beta\, ds \tag{1}$$

where n_α is the unit normal vector to the surface element, ds, of the deformed configuration, $\underset{\sim}{g}_\beta$ is the base vector of x'-y'-z' coordinate system,
and $\sigma_{\alpha\beta}$ are the components of Cauchy stress tensor or the true stress tensor. Similarly, when referring to the reference configuration at $t=t_0$, the

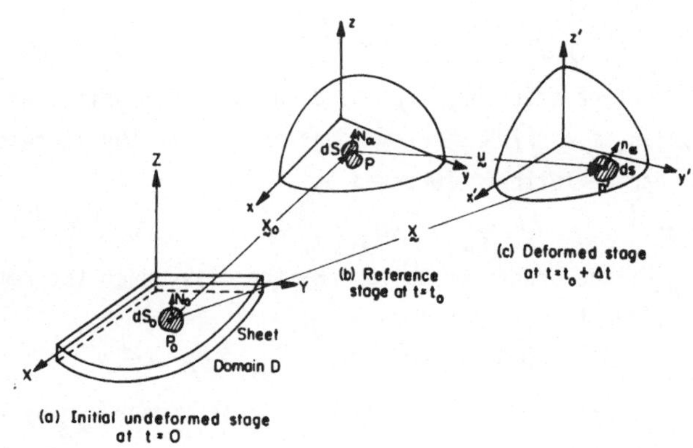

Figure 1   Kinematics of deformation of sheet element at various
stages.

actual force vector, $d\underset{\sim}{F}$, becomes

$$d\underset{\sim}{F} = t_{\alpha\beta} N_\alpha \underset{\sim}{g}_\beta \, dS \tag{2}$$

where N_α is the unit normal vector to the surface element dS of the reference configuration, and $t_{\alpha\beta}$ are the components of the first kind of Piola-Kirchhoff stress tensor, or nominal stress tensor. Since this stress tensor has an inherent non-symmetric characteristic, it is awkward to use in a constitutive equation with a symmetric strain tensor. Alternatively, the actual force, $d\underset{\sim}{F}$, when referred to the reference configuration at $t=t_0$, can also be expressed as

$$d\underset{\sim}{F} = (u_{\beta,\gamma} + \delta_{\beta\gamma}) S_{\alpha\gamma} N_\alpha \underset{\sim}{g}_\beta \, dS \tag{3}$$

where $u_{\beta,\gamma}$ denotes derivative of the displacement vector $\underset{\sim}{u}$, $\delta_{\beta\gamma}$ is the Kronecker delta, and $S_{\alpha\gamma}$ are the components of the second kind of Piola-Kirchhoff stress tensor, which is symmetric and more suitable to use in a stress-strain law.

The relationships between the three stress tensors may be written as

$$J\sigma_{\alpha\beta} = x_{\alpha,\gamma} \, t_{\beta\gamma} = x_{\alpha,\gamma} \, x_{\beta,\delta} \, S_{\gamma\delta} \tag{4}$$

where $x_{\alpha,\gamma}$ and $x_{\beta,\delta}$ are deformation gradients, and J is the determinant of the deformation gradient.

If a virtual displacement vector, $\delta\underset{\sim}{u}$, is imposed onto a material particle, at time $t=t_0 + \Delta t$, the virtual work δw reads, with respect to the reference configuration,

$$\delta w = \int_S \underset{\sim}{f} \cdot \delta\underset{\sim}{u} \, dS \tag{5}$$

where $\underset{\sim}{f}$ is the surface traction, expressed in force per unit area of reference configuration, $\underset{\sim}{f} = d\underset{\sim}{F}/dS$. From equation (2) and the expression $\delta\underset{\sim}{u} = \delta u_\gamma \underset{\sim}{g}_\gamma$, eq. (5) can be further expressed as

$$\delta w = \int (t_{\alpha\beta} N_\alpha \underset{\sim}{g}_\beta) \cdot (\delta u_\gamma \underset{\sim}{g}_\gamma) \, dS = \int t_{\alpha\beta} N_\alpha \, \delta u_\beta \, dS \tag{6}$$

since $\underset{\sim}{g}_\beta \cdot \underset{\sim}{g}_\gamma = \delta_{\beta\gamma}$; the kronecker delta. By the divergence theorem and since $t_{\alpha\beta,\alpha} = 0$, equation (6) becomes.

$$\delta w = \int_V (t_{\alpha\beta} \, \delta u_\beta)_{,\alpha} \, dV = \int t_{\alpha\beta} \, \delta u_{\beta,\alpha} \, dV \tag{7}$$

Replacing the first kind of Piola-Kirchhoff stress tensor $t_{\alpha\beta}$ by the second kind in eq. (7), we obtain

$$\delta w = \int (u_{\beta,\gamma} + \delta_{\beta\gamma}) S_{\alpha\gamma} \, \delta u_{\beta,\alpha} \, dV$$

$$= \frac{1}{2} \int S_{\alpha\gamma} \, \delta (u_{\gamma,\alpha} + u_{\alpha,\gamma} + u_{\beta,\gamma} u_{\beta,\alpha} + \delta_{\alpha\gamma}) \, dV \tag{8}$$

Introducing the incremental Lagrangian strain tensor

$$dE_{\alpha\beta} = \frac{1}{2}\left(u_{\alpha,\beta} + u_{\beta,\alpha} + u_{\gamma,\alpha}\,u_{\gamma,\beta}\right) \tag{9}$$

in eq. (8) and with eq. (5), we have

$$\delta w = \int_S \underset{\sim}{f}\cdot\delta\underset{\sim}{u}\,dS = \int S_{\alpha\beta}\,\delta(dE_{\alpha\beta})\,dV \tag{10}$$

Equation (10) is the principle of virtual work.

3 Yield function and Flow Rules

We assume that Hill's yield function [13] of Cauchy stresses can be extended to that of Piola-Kirchhoff stresses, then the yield function f, under
the plane-stress condition and with normal anisotroy R, becomes

$$f(S_{ij}) = S_x^2 + S_y^2 - \frac{2R}{1+R}\,S_x S_y + \frac{2(1+2R)}{1+R}\,S_{xy}^2 . \tag{11}$$

We further assume that the flow rules are given by

$$dE_{ij} = \frac{\partial f}{\partial S_{ij}}\,d\lambda . \tag{12}$$

Introducing the materials property in the plastic state as $f(S_{ij}) = \bar{S}^2$, the
corresponding effective incremental strain $d\bar{E}$ is defined according to the
condition that $S_{ij}\,dE_{ij} = \bar{S}\,d\bar{E}$. This results in

$$d\bar{E} = \frac{1+R}{\sqrt{1+2R}}\,\sqrt{dE_x^2 + dE_y^2 + \frac{2R}{1+R}\,dE_x dE_y + \frac{2}{1+R}\,dE_{xy}^2} . \tag{13}$$

Then, it can be shown that $S_{\alpha\beta}\,\delta(dE_{\alpha\beta}) = \bar{S}\,\delta(d\bar{E})$, and eq. (10) is expressed
as

$$\delta\phi = \int_V \bar{S}\,\delta(d\bar{E})\,dV - \int_S \underset{\sim}{f}\cdot\delta\underset{\sim}{u}\,dS = 0 \tag{14}$$

Equation (14) is the variational form of the state equation which is the
basis of the finite element discretization.

In order to define the material's property completely, $\bar{S}$ must be determined
from the true stress $(\bar{\sigma})$ and true strain $(\bar{\epsilon})$ curve obtained by an uniaxial
tensile test. With reference to the stress-strain curves shown in Fig. 2,
it is assumed that the energy dissipation during an incremental deformation
are the same in the two expressions. The relationship is approximately
given by

$$(\bar{\sigma} + \bar{\sigma}_0)\,d\bar{\epsilon} = (\bar{S} + \bar{S}_0)\,d\bar{E} \tag{15}$$

where $\bar{\sigma} = \bar{\sigma}_0 + h_0 d\bar{\epsilon}$ and $\bar{S} = \bar{S}_0 + H_0 d\bar{E}$, with $h_0 = d\bar{\sigma}/d\bar{\epsilon}$ and $H_0 = d\bar{S}/d\bar{E}$
evaluated at $t = t_0$.
Noting that $d\bar{\epsilon} = \ln(1 + \frac{d\ell}{\ell_0})$ amd $d\bar{E} = \frac{d\ell}{\ell_0} + \frac{1}{2}(d\ell/\ell_0)^2$, where $d\ell/\ell_0$ is the
nominal strain in the uniaxial tension test, we obtain

$$\bar{S} = \bar{\sigma}_o + (h_o - 2\bar{\sigma}_o)\, d\bar{E} \qquad\qquad (16)$$

since $S_o = \bar{\sigma}_0$ at $t = t_0$. This is the stress-strain equation to be used in
the computing procedure. It is to be noted that the effective true strain
increment is related to Lagrangian strain increment as

$$d\bar{\varepsilon} = \frac{1}{2} \ln(1 + 2\,d\bar{E}) , \qquad\qquad (17)$$

so that the current total effective true strain can be updated according to
$\bar{\varepsilon} = \bar{\varepsilon}_o + d\bar{\varepsilon}$, where $\bar{\varepsilon}_o$ is the total strain accumulated up to $t = t_0$.

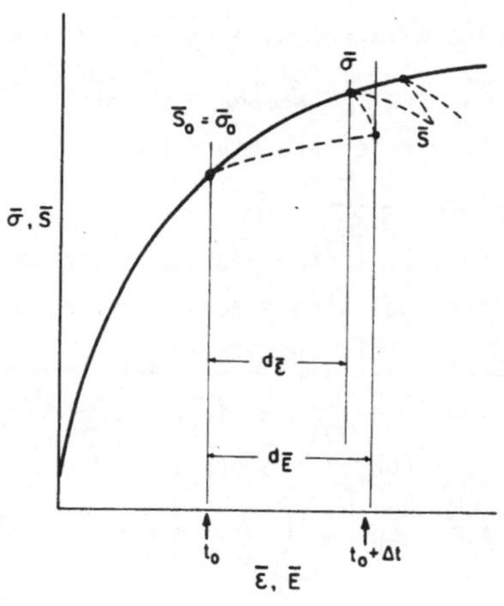

Figure 2   Schematic representation of the stress-strain relationships.

4   Finite Element Discretization

Let the sheet metal be decomposed into an assemblage of triangular elements
and a local Cartesian coordinate system is considered for an individual
element. The displacement field within the element is chosen to be u, v, w
and are assumed to vary linearly within that element. For a typical element, the nodal point displacements are denoted u_i, v_i, and w_i where i=1,
2, 3 as shown in Fig. 3. Then the displacement field within the element is

represented by

$$\left\{ \begin{matrix} u \\ v \\ w \end{matrix} \right\} = \underline{N}\, \underaccent{\sim}{u} \quad , \quad \underaccent{\sim}{u}^{T} = \left\{ u_1, v_1, w_1, u_2, v_2, w_2, u_3, v_3, w_3 \right\} , \tag{18}$$

where $\underline{N}$ is the linear interpolation function matrix, and $\underaccent{\sim}{u}$ is the nodal point velocity vector, superscript T denoting transpose.

Applying the differential operator to eq. (18), the Lagrangian strain components are given by

$$d\underaccent{\sim}{E} = \underline{B}\, \underaccent{\sim}{u} \quad , \quad d\underaccent{\sim}{E}^{T} = \left\{ dE_x, \; dE_y, \; dE_{xy} \right\} , \tag{19}$$

according to

$$dE_x = \frac{\partial u}{\partial x} + \frac{1}{2} \left\{ \left(\frac{\partial u}{\partial x}\right)^2 + \left(\frac{\partial v}{\partial x}\right)^2 + \left(\frac{\partial w}{\partial x}\right)^2 \right\}$$

$$dE_y = \frac{\partial v}{\partial y} + \frac{1}{2} \left\{ \left(\frac{\partial u}{\partial y}\right)^2 + \left(\frac{\partial v}{\partial y}\right)^2 + \left(\frac{\partial w}{\partial y}\right)^2 \right\}$$

$$dE_{xy} = \frac{1}{2} \left\{ \frac{\partial u}{\partial y} + \frac{\partial v}{\partial x} + \frac{\partial u}{\partial x}\frac{\partial u}{\partial y} + \frac{\partial v}{\partial x}\frac{\partial v}{\partial y} + \frac{\partial w}{\partial x}\frac{\partial w}{\partial y} \right\}.$$

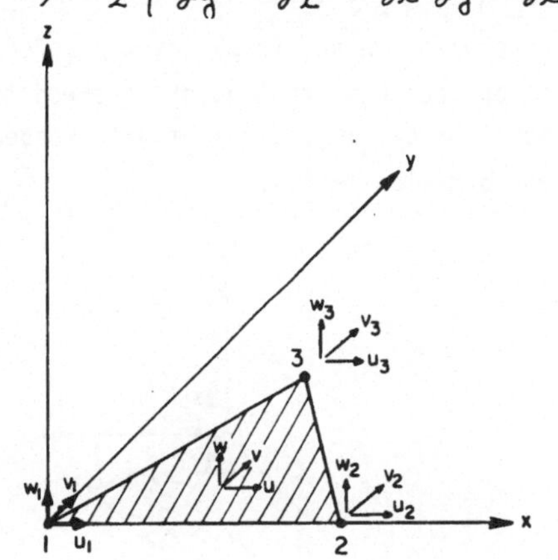

Figure 3   Local coordinates of a triangular element with nodal displacements u_i, v_i and w_i (i = 1, 2, 3).

The strain-displacement matrix $\underline{B}$ in eq. (19) contains linear and nonlinear terms and is a function of $\underaccent{\sim}{u}$. The effective strain increment $d\bar{E}$ given by eq. (13) is then expressed in the vector form, by

$$d\bar{E} = \left(\frac{2}{3}\, d\underaccent{\sim}{E}^{T} \underline{D}\, d\underaccent{\sim}{E} \right)^{1/2} \tag{20}$$

where

$$\underline{D} = \frac{3}{2}\frac{1+R}{1+2R} \begin{bmatrix} 1+R & R & 0 \\ R & 1+R & 0 \\ 0 & 0 & 2 \end{bmatrix}$$

Using eqs. (18), (19) and (20), the discretized form of eq. (14) for the m-th element becomes,

$$\delta\phi^{(m)} = \int \frac{1}{3}\frac{\bar{S}}{dE}\,\delta\underset{\sim}{u}^T\underline{A}^T\underline{D}\,\underline{B}\,\underset{\sim}{u}\,dV - \int \delta\underset{\sim}{u}^T\underline{N}^T\underset{\sim}{f}\,dS \;, \qquad (21)$$

where $\underline{A}$ is the matrix resulting from the derivative of $\underline{B}\,\underset{\sim}{u}$ with respect to $\underset{\sim}{u}$, and its components are functions of $\underset{\sim}{u}$. Equation (21) has to be transformed into the global coordinate system in which the element equations are assembled.

The coordinate transformation is achieved in the following manner: Let (X,Y,Z) with base vector $\underset{\sim}{e}_i$ and (x,y,z) with base vactor $\underset{\sim}{G}_i$ (i = 1, 2, 3) be the global and local Cartesian coordinate systems, respectively (Fig. 4). The transformation of the local coordinates into the global coordinates can be expressed by

$$\underset{\sim}{x} = \underline{\Lambda}\{\underset{\sim}{X} - \underset{\sim}{X}_o\} \;, \qquad (22)$$

where $\underset{\sim}{x}$ is the position vector of an arbitrary point P referred to (x, y, z) system, $\underset{\sim}{X}$ is the position vector of that point referred to (X, Y, Z) system; $\underset{\sim}{X}_o$ denotes the position of (x,y,z) system with respect to (X, Y, Z) system, and $\underline{\Lambda}$ is the transformation matrix.

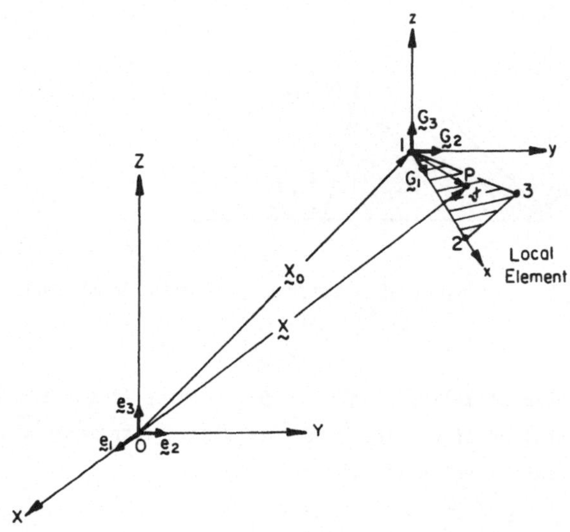

Figure 4   Local and global Cartesian coordinate systems.

The transformation matrix $\underline{\lambda}$ consists of direction cosines as

$$\underline{\lambda} = \begin{bmatrix} \underline{G}_1 \cdot \underline{e}_1 & \underline{G}_1 \cdot \underline{e}_2 & \underline{G}_1 \cdot \underline{e}_3 \\ \underline{G}_2 \cdot \underline{e}_1 & \underline{G}_2 \cdot \underline{e}_2 & \underline{G}_2 \cdot \underline{e}_3 \\ \underline{G}_3 \cdot \underline{e}_1 & \underline{G}_3 \cdot \underline{e}_2 & \underline{G}_3 \cdot \underline{e}_3 \end{bmatrix} \tag{23}$$

The base vectors of the (x, y, z) coordinate system are expressed in terms of the global coordinate system as

$$\underline{G}_1 = \frac{\underline{R}_{12}}{|\underline{R}_{12}|} \; , \quad \underline{G}_3 = \frac{\underline{R}_{12} \times \underline{R}_{13}}{|\underline{R}_{12} \times \underline{R}_{13}|} \; , \quad \underline{G}_2 = \underline{G}_3 \times \underline{G}_1 \tag{24}$$

where $\underline{R}_{12}$, $\underline{R}_{13}$ are position vectors of nodal points 2 and 3, respectively, and given by

$$\underline{R}_{12} = (X_2 - X_1)\underline{e}_1 + (Y_2 - Y_1)\underline{e}_2 + (Z_2 - Z_1)\underline{e}_3 \; ,$$
$$\underline{R}_{13} = (X_3 - X_1)\underline{e}_1 + (Y_3 - Y_1)\underline{e}_2 + (Z_3 - Z_1)\underline{e}_3 \; .$$

Similarly, the local-global displacement transformation equation for each node, is

$$\begin{Bmatrix} u \\ v \\ w \end{Bmatrix} = \underline{\lambda} \begin{Bmatrix} U \\ V \\ W \end{Bmatrix} \tag{25}$$

and

$$\underline{u} = \underline{\Lambda} \; \underline{U}$$

where $\underline{\Lambda}$ is the 9X9 matrix whose component is $\underline{\lambda}$.

Element equation (21) is now transformed into the global coordinate system and assembled to give

$$\delta \phi = \sum_{m=1}^{M} \left\{ \int \frac{2}{3} \frac{\bar{S}}{dE} \delta \underline{U}^T \underline{\Lambda}^T \underline{A}^T \underline{D} \underline{B} \underline{\Lambda} \underline{U} dV \right.$$
$$\left. - \int \delta \underline{U}^T \underline{\Lambda}^T \underline{N}^T \underline{f} dS \right\} = 0 \tag{26}$$

Note that $\underline{u}$ in $\underline{A}$ is now expressed in terms of $\underline{U}$.
Due to the arbitrariness of $\delta \underline{U}$, eq. (26) results in the system of equations to be solved for $\underline{U}$ under the appropriate boundary conditions.
The numerical solution procedure and further detail of the formulations can be found in Reference [12]

5   Applications

Two applications are presented. One is stretching of a circular blank and the other is stretching of a rectangular strip both by a punch with hemispherical head. The results of punch stretching of a circlar blank were compared with the axisymmetric solutions previously reported [6], and the results of stretching of a rectangular strip were compared with the experimental results found in literature [15].

Boundary conditions:

The boundary conditions consist of the geometric boundary conditions and the traction boundary conditions. Geometrically, in the regions where the punch and sheet are in contact, the positions of nodal points can be described by a mathematical expression such that they are moving in accordance with the geometry of the punch profile. The configuration of nodal point movement is shown in Fig. 5. This geometrical relation, when expressed in terms of global coordinate system, gives

$$(X+U)^2 + (Y+V)^2 + (Z+W+C)^2 = r_p^2,$$

where X, Y and Z are current coordinates of nodal points with respect to global coordinate system; U, V and W are incremental displacements and C is a parameter which can be expressed as

$$C = r_p - h,$$

with h as the current punch depth and r_p as the punch radius.

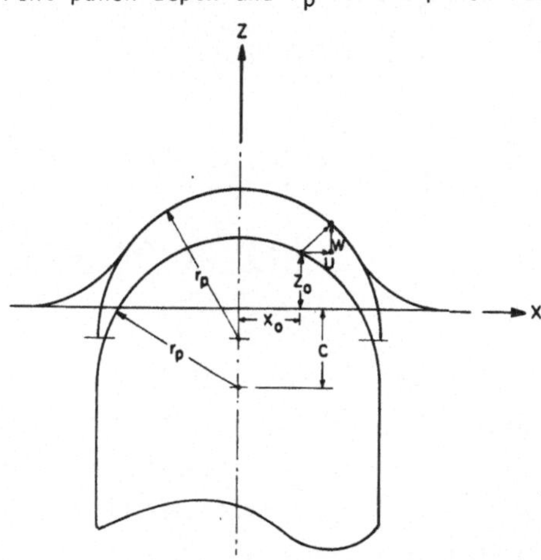

Figure 5   Geometrical requirement for the node in contact with the punch head.

Since the solution procedure uses the Newton-Raphson method, the geometrical boundary condition must be linearized and expressed in terms of the displacement perturbations [12].

The traction boundary conditions associated with the nodal points which are in contact with the punch profile can be formulated with reference to Fig. 6. Two sets of coordinate systems, (X, Y, Z) and (X', Y', Z') are considered. The force acting at a contact point is F_n, the projection of the frictional force F_t on the plane X' - Z' is T and F't is the projection of F_t on the X - Y plane. It can be shown that the components of nodal-point force associated with a contact point are given by

$$F_x = F_n \sin\theta_1 \cos\theta_2 - \frac{\cos\theta_1 \cos\theta_3}{A \cos(\theta_3 - \theta_2)} F_t \; ,$$

$$F_Y = F_n \sin\theta_1 \sin\theta_2 - \frac{\cos\theta_1 \sin\theta_3}{A \cos(\theta_3 - \theta_2)} F_t \; ,$$

$$F_Z = F_n \cos\theta_1 + \frac{\sin\theta_1}{A} F_t \; ,$$

where

$$A = \sqrt{\left\{ \frac{\cos\theta_1 \cos\theta_3}{\cos(\theta_3 - \theta_2)} \right\}^2 + \left\{ \frac{\cos\theta_1 \sin\theta_3}{\cos(\theta_3 - \theta_2)} \right\}^2 + \sin^2\theta_1} \; .$$

The angles θ_1, θ_2 and θ_3 are expressed in terms of the coordinates of a contact point (X, Y, Z) and the displacements of the point (U, V, W).

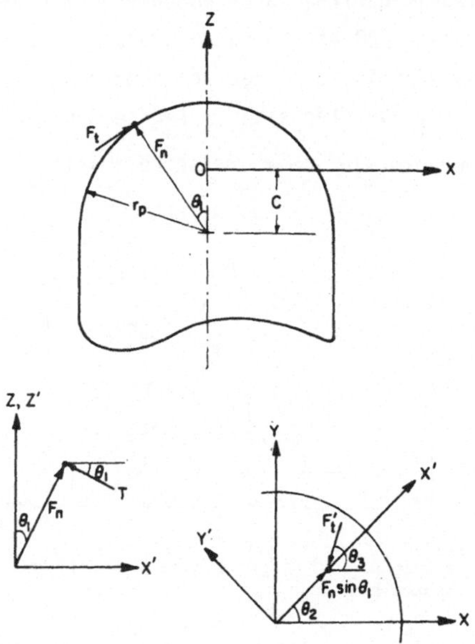

Figure 6   Tractional boundary condition for the node in contact with the punch head.

These relationships are prescribed into the nodal-point load vector, $\underset{\sim}{F}$, whenever a nodal point comes in contact with the punch head. In the present formulation, the frictional force F_t, is set equal to μF_n where μ is the coefficient of friction.

<u>Punch stretching of a circular blank:</u>

The basic material and process parameters used in the computation are listed as follows [6]:

    material: 2036-T4 Aluminum alloy

    stress-strain characteristic: $\bar{\sigma} = 592\ (\bar{\varepsilon})^{0.222}$ N/mm^2

    R value: 0.685

    material thickness: 1.27 mm

    radius of punch: 19.05 mm

    radius of die opening: 20.32 mm

    coefficient of friction, μ : 0.2 and 0.

A sector of sheet metal ($\theta = 9$ degree) with the boundary conditions and the mesh system (shown schematically) used for the finite element analysis is shown in Fig. 7.

The deformation step is controlled by punch head increment which is assigned a value about 1/20 of the punch radius. If, for a particular increment, the step size is too large and some nodal points have moved far inside the punch head, the step size is reduced so that these points are located on the punch profile. With reasonable initial guess for the dis-

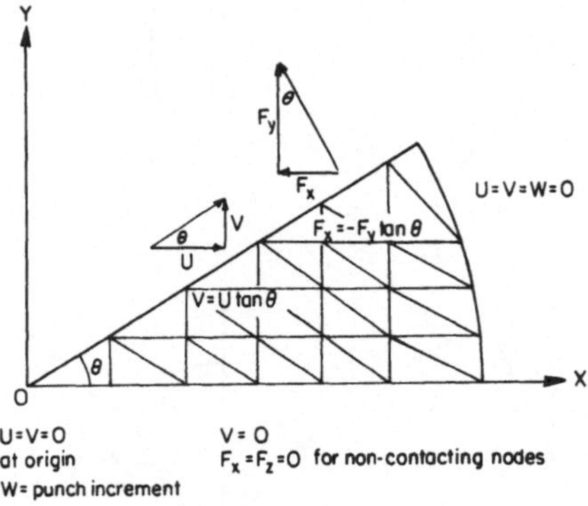

Figure 7   A sector of a circular blank with the geometrical and tractional boundary conditions and the mesh system.

placement increments and using the initial value of deceleration coefficient as 0.2, the solution generally converged within 10 iterations for a
single step with the fractional norm of less than 0.00005.

The results are shown in terms of the thickness strain distributions for
various punch penetrations for the two values of coefficient of friction in
Figs. 8 and 9. Also shown in the figures are comparisons with the axisymmetric finite element solutions. Considering the effects of element
arrantgements and total number of elements on the solutions, agreement between the two solutions is satisfactory.

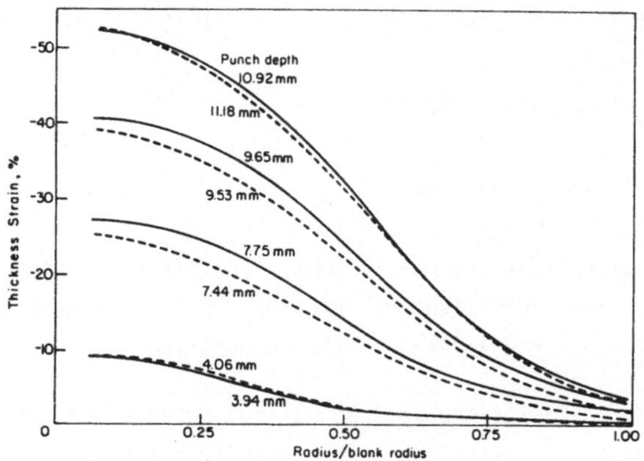

Figure 8   Comparison of the present results (solid curves) with axisymmetric solutions [6] (dashed curves) for thickness strain distribution ($\mu = 0$).

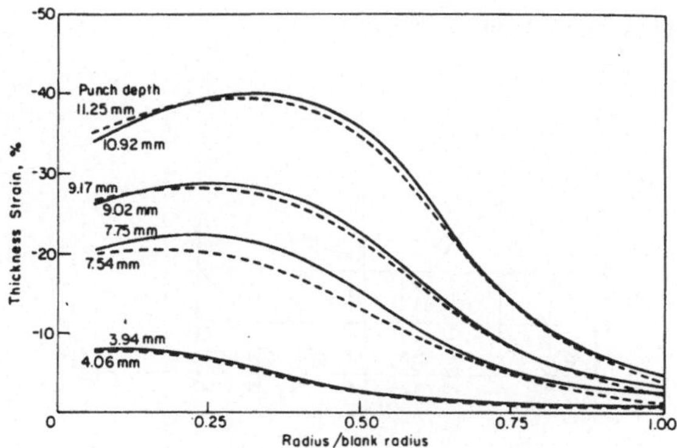

Figure 9   Comparison of the present results (solid curve) with axisymmetric solution [6] (dashed curves) for thickness strain distribution ($\mu = 0.2$).

<u>Punch stretching of a rectangular strip:</u>

The Nakazinna method [14] is one of the test methods for determining the
forming limit curves of sheet materials. It uses a hemispherical punch
with rectangular strips with different blank widths to vary the straining
modes of test specimens. Hecker [15] used the method to determine the
forming limit curves of aluminum alloys and aluminum-killed steels.

Simulation of the test is performed using the following conditions:

        material: 2036-T4 Aluminum alloy
        stress-strain characteristic: $\bar{\sigma} = 592\ (\bar{\epsilon})^{0.245}$ N/mm^2
        R value: 0.78
        material thickness: 1.02 mm
        material width: 50.8 mm
        material length: 203.2 mm
        punch radius: 50.8 mm

Two sets of computations are carried out. One for frictionless condition
and the other, with coefficient of friction μ = 0.2. Figure 10 illustrates
the portion of sheet domain used in the computation.

The thickness strain distributions at various punch depths are shown in
Fig. 11. The effect of friction can be seen in the strain distributions in
the region where the strip is in contact with the punch. However, the
location of the critical site for failure is not affected by the friction
condition, because maximum thinning occurs at the element 21 (see Fig. 10)
for both friction conditions.

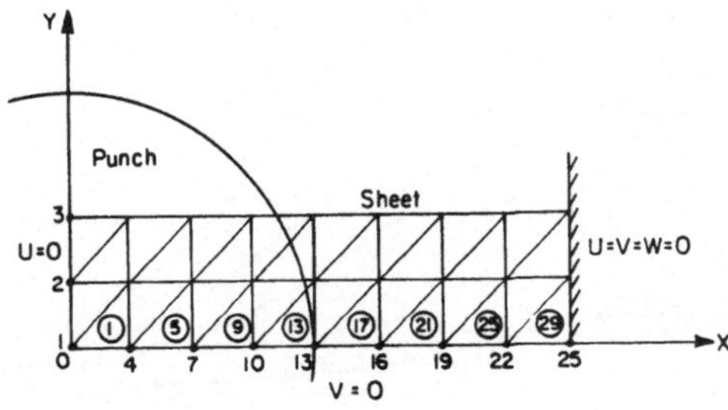

Figure 10   The rectangular strip domain used in punch stretching simula-
             tion with the boundary conditions.

The strain path for the element 21 for μ = 0 was computed and compared with
that by experiment for the lubricated case in Fig 12. The forming limit
curve shown in the figure was determined from the experiments. It is seen
that the experimental and theoretical results agree to each other very
well. The similar comparison is shown for the dry condition in Fig. 13.
It indicates that the experimental strain path agrees very well with that

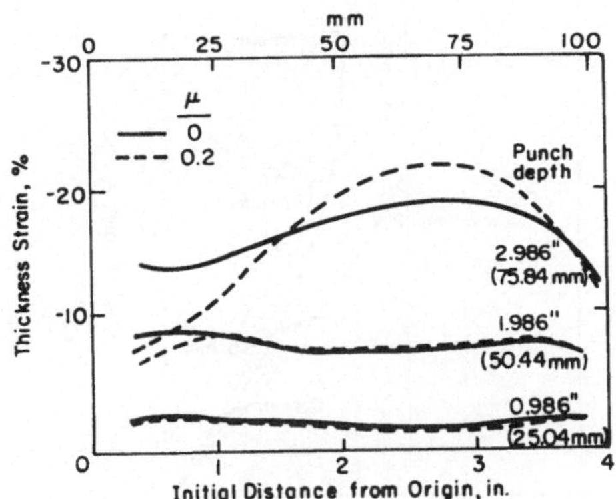

Figure 11   Thickness strain distributions in the rectangular strip at
various stages of punch penetration.

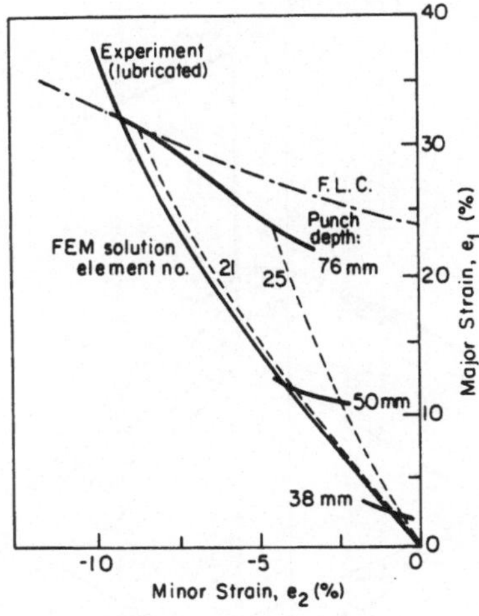

Figure 12   Comparison of strain path of the critical site predicted by the
analysis (μ = 0) with the experiment for the lubricated case
in rectangular strip stretching.

for the element 25. However, the computed results show the element 21 as a critical site. This may indicate that the coefficient of friction corresponding to the dry condition is larger than 0.2. Finally, Fig. 14 shows an example of geometry of the stretched rectangular strip for μ = 0.

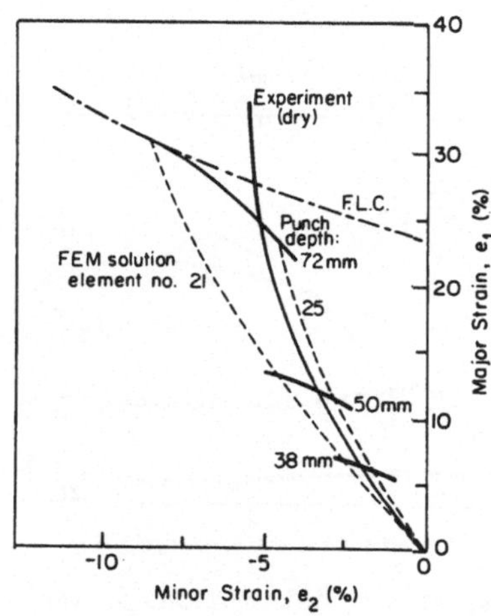

Figure 13   Comparison of strain paths predicted by the analysis (μ = 0.2) with the experimental result under the dry condition in rectangular strip stretching.

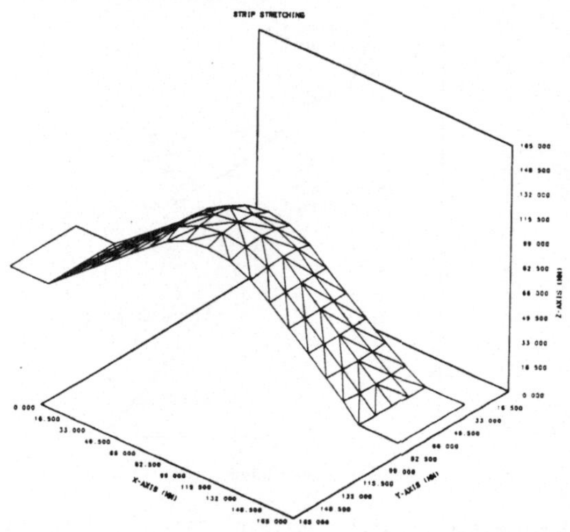

Figure 14   Geometry of the stretched rectangular strip (μ = 0).

6 Conclusions

The principle of virtual work was established, taking into account the finite deformation, and the discretization of the finite element domain was achieved first at the elemental level. The local elemental quantities were transformed into the global quantities. Then the element equations were assembled, and the geometrical and tractional boundary conditions were introduced to the final global equations.

The results of the analysis of punch stretching of a circular blank with a hemispherical-head punch showed excellent agreement with the axisymmetric finite element solutions. This confirmed the validity of the present finite element code. The solutions were obtained for non-symmetric problem of punch stretching of a rectangular strip. It was demonstrated that the solutions were reasonable even in detail in comparison with some experimental observations. Thus it is concluded that the present rigid-plastic finite element method expands its analysis capability of sheet metal forming processes to forming of general shapes.

Acknowledgements

The authors wish to thank Mrs. Carmen Marshall for preparing the manuscript.

References

[1]   Zienkiewicz, O. C.; Onate, E; Heinrich, J. C.: Plastic Flow in Metal Forming, I - Coupled Thermal, II - Thin Sheet Forming, Appl. of Num. Methods to Forming Processes, ASME, AMD, 28 (1978) 107-120.

[2]   Lee, S. H.; Kobayashi, S.: Rigid-Plastic Analysis of Bore Expanding and Flange Drawing with Anisotropic Sheet Metals by the Matrix Method, Proc. 15th Int. Mach. Tool Des. & Res. Conf. (1974) 561-569.

[3]   Wifi, A. S.: An Incremental Complete Solution of the Stretch-Forming and Deep-Drawing of a Circular Blank Using a Hemispherical Punch, Int. J. Mech. Sci., 18 (1976) 23-31.

[4]   Mehta, H. S; Kobayashi, S.: Finite Element Analysis and Experimental Investigation of Sheet Metal Stretching, J. of Eng. for Ind., Trans. ASME, 95 (1973) 874-880.

[5] Wang, N. M.; Budiansky, B.: Analysis of Sheet Metal Stamping by a Finite-Element Method, J. of Appl. Mech., Trans. of ASME, 45 (1978) 73-82.

[6] Kim, J. H; Oh, S. I.; Kobayashi, S.: Analysis of Stretching of Sheet Metals with Hemispherical Punch, Int. J. Mach. Tool Des. Res., 18 (1978) 209-226.

[7] Kaftanoglu, B.; Tekkaya, A. E.: Complete Numerical Solution of the Axisymmetrical Deep-Drawing Problem, J. of Eng. Matr. and Technol., Trans. ASME, 103 (1981) 326-332.

[8] Kobayashi, S.; Kim, J. H.: Deformation Analysis of Axisymmetric Sheet Metal Forming Processes by the Rigid-Plastic Finite Element Method, Mechanics of Sheet Metal Forming - Material Behavior and Deformation Analysis, Plenum Press, New York (1978) 341-365.

[9] Wang, N. M.: Wenner, M. L.: Elastic-Viscoplastic Analysis of Simple Stretch Forming Problems, ibid, 367-402.

[10] Rebelo, N.; Kobayashi, S.: Axisymmetric Punch Stretching of Strain-Rate Sensitive Sheet Metals," Proc. 8th NAMRC, SME (1980) 235-238.

[11] Onate, E.; Zienkiewicz, O. C.: A Viscous Shell Formulation for the Analysis of Thin Sheet Metal Forming, Int. J. Mech. Sci., Vol. 24, (1983) 305-335.

[12] Toh, C. H.: Process Modeling of Sheet Metal Forming of General Shapes by the Finite Element Method Based on Large Strain Formulation, Ph.D. dissertation, Mechanical Engineering, Univ. of Calif., Berkeley, Sept. 1983.

[13] Hill, R.: The Mathematical Theory of Plasticity, Oxford Univ. Press, London (1950).

[14] Nakazima, K; Kikuma, T; Hasuka , K.: Study on the Formability of Steel Sheets, Yawata Technical Report, No. 264 (1968), 141-154.

[15] Hecker, S., S.: A Simple Forming Limit Curve Technique and Results on Aluminum Alloys, General Motors Research Publication GMR-1220 (1972).

A Review of Development and Use of the Upper Bound Approach to Metal Forming Processes

H. Kudo, Department of Mechanical Engineering, Yokohama National University / Japan

Summary

The development and use of the upper bound approach (UBA) are surveyed to indicate on what bases it stands, to what purposes it has been applied and how it has adapted itself to changing technological requirements and advancing computational facilities. The state of the art of and the problems set on UBA are then summarized. Finally the unique parts which UBA should play to continue useful in the future age of the finite element method are suggested.

List of Symbols

f — friction factor defined by $f = \tau_f / k$

k, k_f — flow stresses of workpiece material in simple shear and compression resp.

k_{fm} — mean of k_f within workpiece

m — strain-rate hardening exponent

S_C — work-tool interface

$S_D{}^*$ — assumed internal surface of velocity discontinuity

S_T, S_V — workpiece surfaces where tractions and particle velocities are specified resp.

T_i — vector component of specified traction

V — workpiece volume

V_i — vector component of specified tool velocity

$v_i, \dot{v}_i$ — vector components of actual velocity and acceleration of workpiece particle

$v_i{}^*, \Delta v_t{}^*$ — kinematically admissible velocity component and tangential velocity discontinuity resp.

V_n, v_n — normal components to S_C of tool velocity and workpiece particle velocity resp.

$\dot{W}_C{}^*$ — upper bound to total rate of work being done by tool over S_C

$\dot{W}_d, \dot{W}_f$ — calculated rates of work due to plastic deformation and

$$\dot{W_s}, \dot{W_t}$$ frictional sliding resp.
$\dot{W_s}, \dot{W_t}$ — calculated rates of work due to internal shearing and traction resp.

x_i — i-th axis of Eulerian co-ordinate system

$\bar{\dot{\varepsilon}}^*, \dot{\varepsilon}_{ij}^*$ — kinematically admissible equivalent strain-rate and component of strain-rate tensor resp.

$\int d\bar{\varepsilon}$ — actual equivalent strain $= \int \bar{\dot{\varepsilon}}\, dt$

$\bar{\varepsilon}_m^*$ — average of $\bar{\dot{\varepsilon}}^*$ within workpiece

θ — current temperature of workpiece particle

μ — coefficient of friction

ρ — density of workpiece material

$\bar{\sigma}$ — equivalent stress due to von Mises

τ_f — frictional stress

0   <u>Introduction and Concept of the Upper Bound Approach (UBA)</u>

It is 32 years since the upper and lower bound theorems were
first applied by Hill to metal forming processes. The use of
UBA has been much in vogue for the last twenty years for the
understanding and analysis of various forming processes, for
the planning and/or control of toolings and operations in metal
forming production and, occasionally, for the production of
technical papers.

Meanwhile stretch of the upper bound theorems in favour of the
authors including the present author or misinterpretation of
the theorems has been frequently made without paying due attention to some limitations inherent in UBA. On the other hand,
due to rapid progress of the finite element method of analysis
(FEM), the question has often been set up recently by researchers
and engineers working in the metal forming field whether UBA
will continue to be useful or it should be wastebasketted.

In the present review paper, the author, by recolleting the history of development of the relevant theorems, the problems to
which UBA was applied and the computational procedures or techniques, tries to point out the limitations that are proper to
UBA and the precautions to be taken when using UBA and then to
answer the above question regarding the future of UBA about

which he is quite optimistic.

The concept of the 'upper bound approach' (UBA) as applied to
Lévy-Mises' materials is illustrated by the flow chart shown in
Fig. 1. The meanings of the symbols in the chart are given in
preceding 'List of Symbols'. If minimization of the upper bound
$\dot{W}_C^*$ to the total rate of work being done by the tool over S_C is
not attempted and only an upper bound $\dot{W}_C^*$ is presented based on
a selected kinematically admissible velocity v_i^*, the result
should, in the opinion of the present author, be called an 'upper bound solution' and the method 'the upper bound method' but
not UBA. The present author does not know if the godfather of
the name 'UBA', W. Johnson, agrees to these definitions. Another confusing term is the 'energy method', which calculates the
total deformation energy ignoring the specified boundary conditions for the actual deformation and, in some cases, the associated redundant work. The resulting rate of energy dissipation
can not then be an upper bound. The term 'bounding method' or
'limit analysis' may be used only when both the upper and lower
bounds are presented.

1　The Medieval History of UBA

The medieval age of UBA opened in 1936 when Gvozdev [1] first
proved the upper and lower bound theorems for the plastic collapse load of structures. It was succeeded by the modern age
in about 1960 which lasted until the beginning of the 1970's
when application of electronic digital computers began to progress rapidly. Before surveying the work and their implica-
tions, the present author must apologize for that the survey is
rather biassed in favour of the Japanese investigators and,
in particular, of his own, because of a limited number of the
existing literatures and the time available to him.

A number of upper and lower bound theorems together with various
variational principles for elastic-and rigid-plastic materials
were proposed and discussed from the middle of the 1940's to the
beginning of the 50's by Prager, Markov, Hill, Hodge, Greenberg,
Drucker and others. According to them, the upper bound for

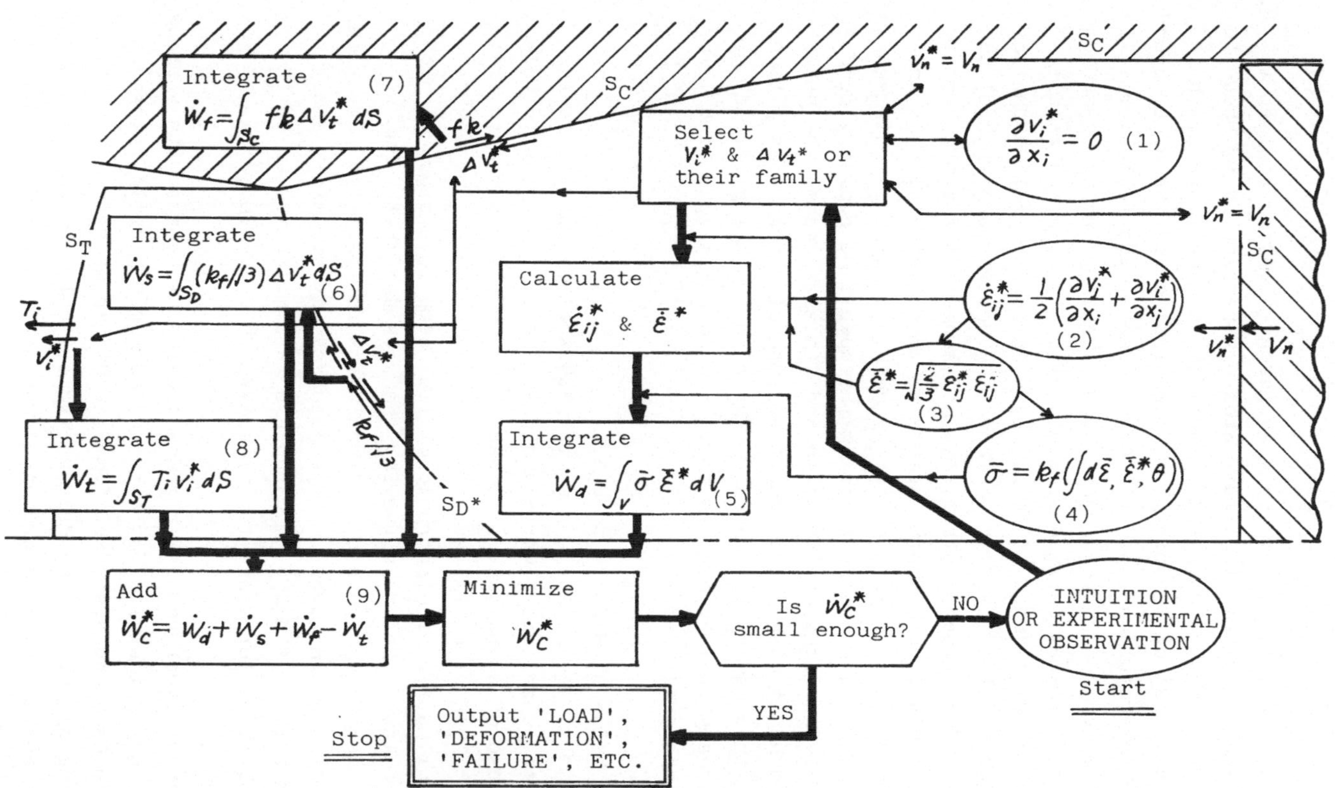

Fig. 1: Concept of the Upper Bound Approach to Metal Forming Process

rigid-plastic von Mises' materials to the actual rate of work at the workpiece surfaces S_V where the velocities V_i are specified is expressed as

$$\int_V k_f \, \bar{\dot{\varepsilon}}^* \, dV + \int_{S_D} (k_f/\sqrt{3}) \Delta V_t^* \, dS - \int_{S_T} T_i \, v_i^* \, dS \geq \int_{S_V} T_i \, v_i \, dS \, . \tag{10}$$

The application of the theorems to metal forming problems was first suggested by Hill [2] and Green [3, 4] in 1951 and 52 for a plane-strain drawing and compression of sheet between smooth flat dies. The upper and lower bounds to the load of indentation of a smooth flat-ended punch having a rectangular or circular cross section onto a flat surface of a non-hardening Tresca material were obtained by Shield and Drucker [5, 6] and Levin [7] in 1953 and 55. Thus the three-dimensional plastic deformation problems were treated analytically for the first time with the aid of the variational principles.

Extremum theorems for a rigid-visco-perfectly plastic body that obeys the von Mises yield criterion were given by Prager in 1954, which were generalized by Hill [8] in 1956. By assuming that the 'work function' defined by $E(\dot{\varepsilon}_{ij}) = \int_0^{\varepsilon_{ij}} \sigma_{ij} \, d\dot{\varepsilon}_{ij}$ is everywhere 'convex' and $k_f = k_{f1}(\bar{\dot{\varepsilon}})^m$, the upper bound theorem is written as

$$\frac{k_{f1}}{1+m} \int_V (\bar{\dot{\varepsilon}}^*)^{1+m} \, dV + \frac{m \, k_{f1}}{1+m} \int_V (\bar{\dot{\varepsilon}})^{1+m} \, dV - \int_{S_T} T_i \, v_i^* \, dS \geq \int_{S_V} T_i \, v_i \, dS. \tag{11}$$

The left hand side of the inequality contains an unknown actual equivalent strain rate $\bar{\dot{\varepsilon}}$ and, therefore, upper bounds to $\int_{S_V} T_i \, v_i \, dS$ are not obtainable from v_i^*. Only when $m=0$, i.e. $k_f = k_{f1}$, or if v_i^* coincides with v_i, Inequ. (11) reduces to Inequ. (10). Note that the velocity discontinuity ΔV_t^* is not allowable for viscous materials since this leads to an infinite flow stress. Upper and lower bound theorems for problems containing interfacial friction that obeys the Amonton-Coulomb law were first proposed by Drucker in 1954. Since the rate of frictional work is not calculable without knowing the actual interfacial pressure, his proposal was too safe resulting in too high an upper bound and too low a lower bound.

In 1953, while carrying out an experiment in metallic ring compression between flat dies, the present author found incidentally that the ring diameter markedly expanded with good lubrication while the inner diameter shrank with poor lubrication. The idea crossed his mind that the change in deformation was a compromise between the deformation and frictional energies. Using the so-called 'parallel deformation field' (i.e., $V_r^* = V_r^*(r)$ and $V_z^* = V_z^*(z)$) and the frictional drag $\tau_f = \mu k_f$, he minimized analytically the sum of the rates of deformation and frictional energy and published the result in 1955 [9]. The result was in good agreement with Kunogi's result obtained by the slab method and published in 1954.

At that time he called his method an 'energy method', because he was unaware of the aforementioned variational principles, although he was one of the translators of the Japanese version of Hill's book [10] that had been published in 1954. But if the present author had been aware of them, he would not have dared to use the idea of minimum energy for that problem accompanied with friction and be indifferent to the lower bound theorem. He proceeded to analize the cold backward can extrusion [11] using the same method and the velocity field proposed by Dipper in 1949. Though there was an unsound treatment of Coulomb's friction, the experimentally and analytically determined extrusion pressures and flow patterns showed good agreement (Fig. 2).

From 1958 to 60, a series of papers were published on UBA to various steady and non-steady state processes of forging and extrusion in plane-strain by Kudo [12] and Johnson and his coworkers [13, 14, 15 16] and for axisymmetric processes by

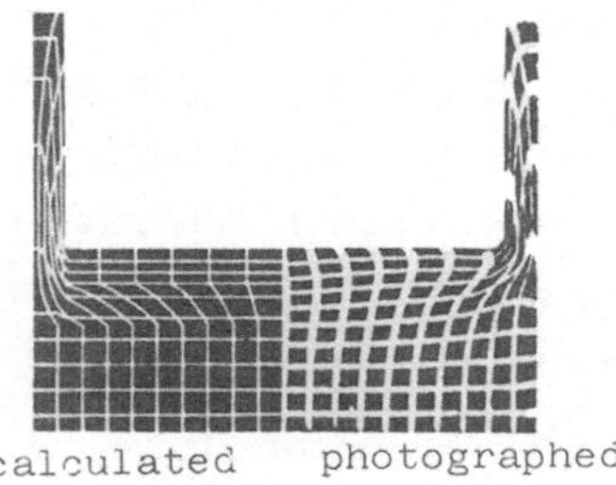

Fig. 2: Comparison of Deformation Patterns Calculated
by UBA and Photographed with Lead Specimen [11]

Kudo [17, 18]. By the use of velocity fields that allowed skin
inclusion, material separation from tool surface or internal
material separation or fracture and resulted in lower upper
bounds than those of sound deformation, Kudo and Johnson even
predicted the occurance of those material defects, see Fig. 3.
The results were generally in good quantitative agreement with
experimental observations [19, 20]. Their velocity fields were
used also to calculate the adiabatic temperature rise [21, 22]
by assuming that the flow stress was unaffected by temperature.
UBA to three-dimensional forming problems seems to have been
made first by Tarnovskii and others [23]. Energy method solutions to three-dimensional shear spinning with an imperfectly
admissible velocity field [24] were also presented.

The kinematically admissible velocity fields used in the above
mentioned literatures were rather simple. The number of free
parameters which express the velocity field and are varied for
the minimization of $\dot{W}_C^*$ was relatively small so as to enable
even analytical minimization. For plane-strain, plastic deformation was regarded as taking place in the form of relative
sliding of a limited number of rigid triangles as first suggested by Hill [2]. For axisymmetry, fields consisting of a few
ring-shaped regions whose meridian sections are rectangular or
triangular were used, these being referred to the 'parallel deformation field'. Kudo proposed, for obtaining low upper bounds
with minimum computation time, to subdivide a workpiece into
several unit rectangular regions. Each region consisted of a
few elements having triangular sections and their configuration
was optimized for various frictional boundary conditions. Those
regions were then assembled so as to fit the given velocity
boundary conditions and their dimensions were optimized to yield
the lowest $\dot{W}_C^*$. This method compares with the
'method of cellular optimization' in the systems engineering.

Other types of velocity field of which the velocity components
are defined by a simple polynomial or trigonometric function of
two spacial co-ordinates to describe barreling of the free side
surface of workpiece in upsetting are found in the monograph due

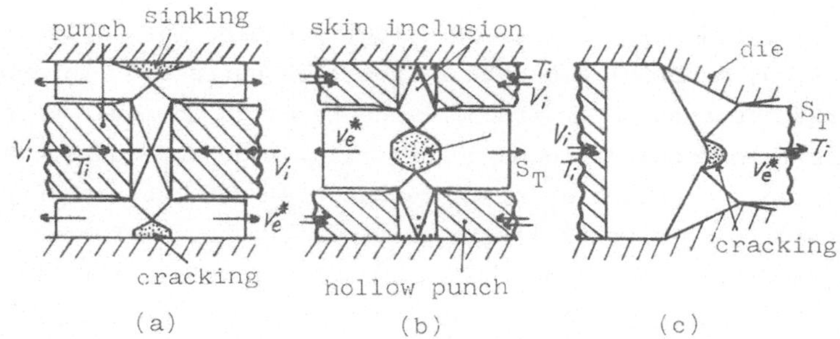

Fig. 3: Predicted Material Defects

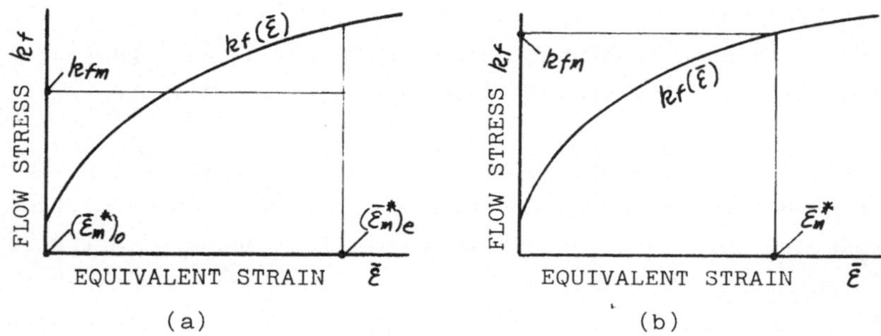

Fig. 4: Determination of Mean Flow Stress

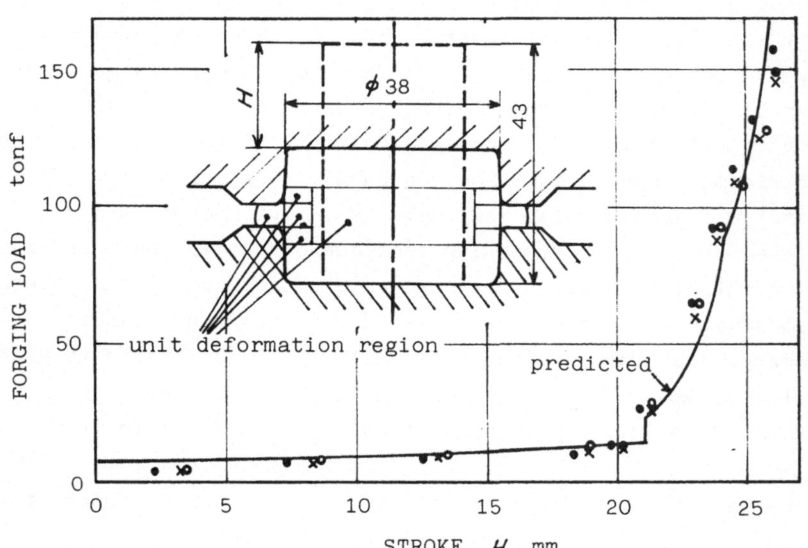

Fig. 5: Comparison of Experimental and UBA Loads of
Hot Semi-Closed Die Forging of Steel [18]

to Tarnovskii et al. [23]. In some cases, they minimized
$k_f \int_V (\bar{\varepsilon}^*)^2 dV$ instead of $k_f \int_V \bar{\varepsilon}^* dV$ to obtain linear equations for
the optimum coefficients.

For work-hardening materials, approximate forming loads were
converted by Kudo [20] from the upper bound loads for a non-
hardening material by using average flow stresses k_{fm} based on
a calculated average equivalent strain $\bar{\varepsilon}_m^* = \int \bar{\varepsilon}_m^* dt$ and an experimental flow stress curve of the material. For steady ex-
trusion processes, this method, due originally to Hill [10],
gave satisfactory results compared with experimental results
(Fig. 4 (a)). For post-steady processes, the modified k_{fm} due
to Kudo (Fig. 4 (b)) resulted in a considerable overestimate of
the forming load, probably because work-softening of material
caused by possible rotation of axes of principal stresses was
not taken into account. When analyzing an axisymmetric hot
semi-closed die forging process, see Fig. 5, Kudo [18] first
conducted UBA for a non-hardening material assuming $f = 1$, and
then determined the average flow stress k_{fm} from a result of a
forging experiment for St 15 Ck. The value k_{fm}=10kgf/mm^2 fit-
ted surprisingly well all other experimental results obtained
with the same workpiece material under different geometrical
conditions.

Meanwhile attempts were made to improve the upper and lower
bound theorems for the Coulomb friction proposed by Drucker.
They are due to Yamada [25], Kudo [17] and Shindo [26]. Kudo
and Takahashi [27] proved in 1964 upper and lower bound theorems with the interfacial friction of shear type, i.e., $\tau_f = f k$
where the constant f denotes the 'friction factor' and k the
'shear flow stress' of material. Their theorem for the upper
bound to the rate of work being done by the tool moving at a
speed of V_i over the material-tool interfaces S_C is written as

$$\int_V k_f \bar{\varepsilon}^* dV + \int_{S_D} (k_f/\sqrt{3}) \Delta v_t^* dS + \int_{S_C} f k \Delta v_i^* dS - \int_{S_T} T_i v_i^* dS \geq \int_{S_D} T_i V_i dS. \quad (12)$$

The use of the friction law, $\tau_f = f k$, may be justified by the
fact that the coefficients of friction μ measured in some forming processes were as nonuniform over the material-tool inter-

face as the measured friction factors f were.

2   The Modern History of UBA

Since 1962, numerous papers on UBA have flown out, only a limited number of which have treated the lower bound methods as well. In them, the advancement has been made principally on the lines of elaborating more complex and flexible velocity fields v_i^* to fit more complicated or irregular geometries and to more precisely simulate the actual material flow for attaining lower upper bounds. To mention some noteworthy examples, the application of the radial or centripetal flow models to the plane-strain and axisymmetric drawing and the extrusion through a wedge-shaped or conical die was made in 1963 by Stepanskii [28] and Avitzur [29, 30]; a hodograph technique for the construction of velocity fields consisting of triangles of parallel flow in axisymmetry was developed in 1965 by Halling and Mitchell [31] and was generalized by Adie and Alexander [32]; Lambert and Kobayashi [33, 34] first introduced in 1967 the stream function method into UBA to plane-strain and axisymmetric steady-state problems and they also proposed superposition of the functions to yield better upper bounds; the use of dual stream function for three-dimensional problems was initiated by Nagpal and Altan [35, 36] in 1975. The method of stream function is, like the above velocity fields consisting of several subdomains, visually plain and is convenient for treating curved die problems, since the stream function is directly derived from the equation of stream line and provides us with incompressible velocity fields.

The expanded application of UBA or the upper bound method based on Inequ. (12), that is only valid for static deformation of perfectly plastic materials whose current surface profiles are exactly known, has been made to those problems in which those premises are violated. In 1964, Halling and Mitchell [37] performed first the optimization of the rigid-triangle velocity field, taking the change in the flow stress along stream lines into consideration for a plane-strain extrusion of a strain-hardening material, though the resulting extrusion pressures did not differ significantly from those determined with the average flow stress k_{fm}. Haddow [38] pointed out that the extrusion pressure thus obtained is not necessarily an upper bound. Even the effects of viscosity and work-hardenability upon internal fracture were discussed with UBA by Thomason [39] and Zimmerman and Avitzur [40].

Extrusion pressures of strain-rate-sensitive materials were approximately calculated first by Avitzur [41] in 1967 using an average flow stress and then by Cristescu [42] in 1975 with a varying flow stress along stream lines. In both cases, the fixed centripetal velocity field having S_D* was used. Upsetting and extrusion of strain and strain-rate affected materials were treated by Lahoti and Altan [43] and by Fenton [44] both in 1975 with simple diversion of the upper bound theorem for the perfectly plastic materials. The temperature distribution due to heat generation and conduction was also incorporated in the former's work. It is likely that those authors had not paid due attention to Hill's Inequ. (11).

The effect of inertia force of the workpiece body upon the working load was approximately assessed by Avitzur et al. [45] in 1972 by merely adding a term for kinetic energy to $\dot{W_c}*$. Tirosh and Kobayashi [46] attempted in 1976 to formulate an upper bound load for the dynamic problem. The expression for the upper bound theorem presented by them contained an unknown quantity $\dot{v_i}$ representing the actual particle acceleration. Examining the effect of $\dot{v_i}$ and replacing it by $\dot{v_i}*$, they made interesting observations on the kinetic effects in some forming processes.

In 1972, Johnson and coworkers [47] presented an upper bound expression for an anisotropic material and they applied it to some plane-strain problems. An upper bound theorem for a porous material was proposed and utilized by Oyane and his coworkers [48] in 1974, in which changes in the density and flow stress were considered. Extrusion of composite materials were also treated with the upper bound method by Avitzur [49].

Osakada et al. [50] were able to predict the process parameters needed to obtain sound products by the use of UBA. Even hydrodynamic lubrication associated with strip rolling was treated courageously with UBA by Avitzur and Grossman [51].

It is quite problematic to apply the existing upper bound theorems to non-steady large deformation processes even when the workpiece material is perfectly plastic since the successive deformation modes are determined by the calculated preceding geometry. This is especially the case for processes, such as ring upsetting [9, 52], open- and closed-die forgings [53, 54, 55, 56] and combined multi-orifice extrusion [12, 17, 32, 57], in which the workpiece has a high degree of freedom of flow. Nevertheless, the successive deformation increments were traced up to a large amount of deformation by many researchers in expectation of a possibility either that the current process geometry would not be far from the actual or that it would not much affect the optimum velocity field for the subsequent step. Fortunately the predicted upper bound loads and the general material flows were found to be in good qualitative and, in many cases, quantitative agreement with the experimental results obtained with non-hardening as well as hardening materials (see, e.g., Fig. 5). It should be noted, however, that in these kinds of processes, the general material flow is largely dependent upon the interfacial frictional stress τ_f that can only be determined from the results of UBA and experiment of the forming process being considered.

The predicted surface geometry of a deformed workpiece sometimes remarkably differed from the actual [58] even if it was obtained by the use of an elaborated velocity field expressed with a number of parameters for optimization especially when the unconstrained surface occupies a considerable portion of the surface of the deforming region of workpiece. Some examples will be seen in Refs. [52, 59] which treat side surface barreling of circular discs or rings under compression between flat rough tools, see Fig. 6.

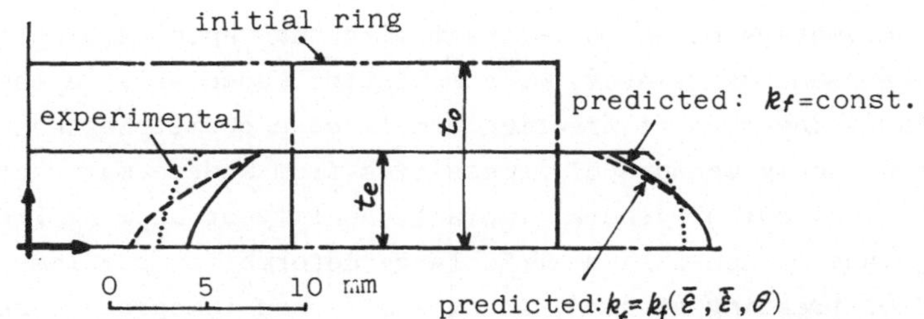

Fig. 6:   Comparison of Experimental and UBA Profiles of
Side Surface in Ring Upsetting [59]

The circumstances are almost the same for steady-state forming
processes, especially when the surfaces of the deforming regions are not contained by tools or rolls. But we find a number of successful research work in which the admissible velocity fields of minimum energy dissipation were selected out
from among a family of fields that are subject to certain limitations imposed empirically. Analyses of chipless forming processes of V-shaped groove on a flat surface due to Kudo and
Tamura [60], those of three-dimensional machining due to Usui
and Masuko [61] and those of sheet and tube spinning processes
due to Hayama and his coworkers [62] are the examples. Usui
and Masuko obtained quantitatively excellent prediction of cut-
ting force and chip flow direction on the bases of the mean
flow stress and the angle of shear plane determined by an orthogonal machining test.

An algorithm to improve the reliability of UBA to steady-state
forming processes that are not entirely contained by tooling
was first proposed by Oh and Kobayashi [63] in 1975 in a study
of the sideways spread in rolling of rectangular bars. It was
certainly persuasive and reduced empirical factors when determining the optimum v_i^*.

3   The Contemporary History of UBA

The contemporary age of UBA history opened early in the 1970's,
when the needs for the use of multi-parameter velocity fields
to cope with large deformation processes with complicated geometry and material properties and the time saving in optimizing

the parameters began to interact with the rapid development of
the hard-and soft-wares for computation especially in the field
of FEM. The need of treating complicated geometries has been
felt not only because of necessities from modern manufacturing
conditions but of general impracticability of appling FEM to
non-plane or non-axisymmetric large deformation problems. The
need of treating strain-, strain-rate- and temperature-sensi-
tive materials has also been raised by industrial practices.

One of the lines to meet the above needs has been the use of
analytical velocity or stream functions that are continuous
throughout or at least in each subdomain of the deforming zone.
The introduction and superposition of stream functions that
were proposed by Lambert and Kobayashi in 1967 [33] opened up a
road to easy construction of admissible velocity function having a number of free parameters and also to simple elimination
of internal velocity discontinuities. The stream function
method was used by, other than the authors of Refs. [43, 44,
59], e.g., Chen and Ling [64], Nagpal et al. [65, 66], Chang and
Choi [67] and Busch [68] for analyzing plane and axisymmetric
deformation in steady as well as non-steady state. The introduction of dual stream function, as is stated above [35, 36],
made it possible to construct velocity fields of flow through
elliptic and rectangular dies. Various optimization procedures were proposed for polynominal velocity functions of high
degree by, e.g., Andresen [69] and Lahoti et al. [70]. For
steady state asymmetric problems, admissible velocity fields to
calculate upper bound solutions were obtained by conformal
transformation by Yang and others [71, 72]. A more intuitive
velocity was proposed by Gunasekera and Hoshino [73]. Kiuchi
and his coworkers [74, 75] have recently developed a general
method by which any cross sectional shapes of die that is
straightly converging could be treated, numerical integration
by computer being needed. But it is to be remarked that in almost all existing velocity fields for asymmetric extrusion and
drawing, the axial velocity was chosen uniform over cross sections having no room for the field to vary depending on the
frictional condition.

Another line to meet the needs stated at the beginning of this
chapter is the use of kinematically admissible velocity fields
consisting of a number of subdomains within which V_i^* are uniform or given by relatively simple functions of spacial coordinates like those used in the early studies. A straight extension of Kudo's unit region method [12] was made by Oudin and
Ravalard for plane-strain processes [76] who added 'unit trape-
zoidal deformation regions' to the previous 'unit rectangular
regions' and presented a programme that was feasible for mini-
computers.

As the number of subdomains increased, the velocities were not
uniquely determined by the subdomain geometry and, therefore,
the minimization was performed by changing the velocities at
the subdomain boundaries as does the rigid-plastic FEM for
forming problems initiated first by Lung and Mahrenholz and Lee
and Kobayashi in 1973, Since 1974, McDermott, Bramley, Avitzur
and others [56, 77, 78, 79] developed step by step programmes
for the computer simulation and design of axisymmetric extrusion and forging processes using subdomains or elements of parallel deformation having rectangular and triangular cross sections, see Fig. 7. This method was called the 'upper bound elemental technique'(UBET). Kiuchi and Shigeta [80] applied it to
asymmetric cold forging. Another unique discretized velocity
field for general three-dimensional problems was proposed by
Gatto and Giarda [81] which consisted of a number of rigid poly-
hedrons.

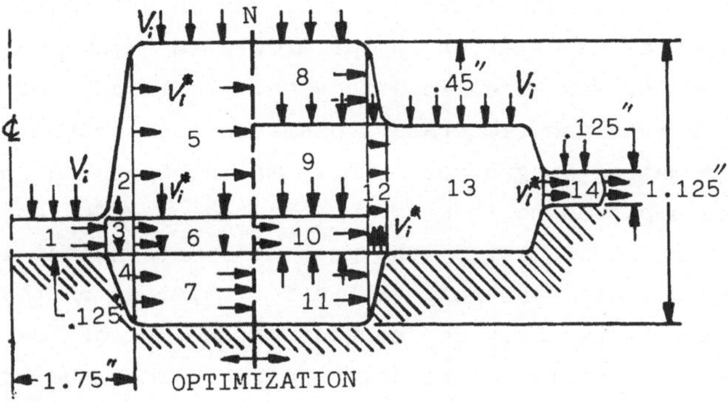

Fig. 7: Subdivision of Workpiece for UBET Treatment

Thus the boundary between UBA and FEM seems to have disappeared and hybrids between them are appearing today. In a modified stream function method due to Tomita et al. [82], the combination of stream functions was optimized employing a descretized form in one dimensional FEM to reduce computing time. Kitahara et al. [83] calculated upper and lower bounds to plane-strain extrusion pressure by a linear programming method using coarser triangular elements than those of FEM.

The results of UBA have been proved from the beginning to be useful in getting ideas for reducing forming loads, controlling material flow and avoiding workpiece defects [12, 15, 16, 17,32, 40]. Indications for the design of forging dies, preforms and operations [55, 77, 84], the elimination of void within workpiece by rolling [85] and the means to correct eccentricity of tubes [75] have also been given recently by using UBA. Kondo and his coworkers have managed to develope a novel cold precision forging process for producing spline pieces of automobile transmission gears based on UBA [86].

4   Present Problems for UBA to Resolve

Imperfect knowledges of the constitutive relations (Equ. (4), Fig. 1) of workpiece materials and the frictional conditions or the f-values on work-tool interfaces still prevent UBA as well as FEM from getting detailed information on the deformation mechanism and metallurgy.

The use of relatively coarse velocity fields in UBA inevitably leads to accumulation of error in the predicted workpiece profile and extent of work-tool interface and the calculated strain, strain-rate and temperature distributions within workpiece as deformation progresses. This implies that the application of the upper bound theorems which hold only for an instantaneous flow subsequent to the past deformation that is exactly known becomes doubtful. It is for this reason that the present author has recently got nervous at the use of the term 'UBA' in large deformation analyses especially when the workpiece surfaces are insufficiently constrained (i.e., $S_C < S_T$)

and is inclining to adopt the rather vague term 'minimum energy method' as he did before [9]. The only way for UBA to overcome this disadvantage may be the use of more sophisticated velocity fields at the expense of increased computational cost amalgamating finally with FEM.

Inability to deriving information on the stress distribution is another sore spot inherent in UBA. Kudo and Shinozaki[88] managed to predict punch pressures in a multi-ram extrusion by regarding one of the work-punch interfaces S_C a traction prescribed surface S_T and utilizing a force balance condition. More general techniques for adding the pressure distribution to the conventional upper bound solution were proposed by Johnson and Mamalis [89] and Kiuchi and Murata [90], although the justification of the methods given in their papers was not persusive enough to the present author.

It is rather surprinsing that the predicted occurance of cracking (Fig. 3) coincided well with experimental results [20, 40] without knowledges of stress when the workpiece material had fully work-hardened. Even the effects of work-hardenability, strain-rate-hardenability and superimposed hydrostatic pressure upon fracture have been taken into account in UBA [40, 91] and qualitatively good results have been obtaind. If a rate of energy dissipation needed to create new surface can be incorporated into UBA, the fracture analysis may be extended to a wider range of work materials.

5   Suggested Parts to be Played by UBA in Future
One of the future function of UBA will be helper of FEM who offers to production and planning engineers preliminary information about forming processes of which geometrical configurations and material properties are so complex as to make FEM computation economically impracticable. The results may be also useful to make subsequent FEM computations cheaper. This function will become useless in the future with further development of computing facilities and FEM.

The more long-lasting application field of UBA will be the inter- and extra-polation of experiences or experimental observations, which drastically reduces computing time and covers shortage of knowledge about the constitutive equation and frictional condition. The velocity field v_i^* may be relatively simple when it is supported by experimental observation. If the average flow stress k_{fm} and friction factor f are obtained in a typical forming experiment by using a deformation model, the loads and modes of flow for similar processes that are predicted with the same k_{fm}, f and model will be sufficiently precise for production purpose even when the velocity field or model adopted is simple [18,61]. In this connection,it is to be remarked that the validity of the in-process combination of a simple plasticity theory with process sensing had been industrially confirmed by the computer control in the strip rolling industry.

The use of experimentally observed current workpiece profile as the base of calculation eliminates the possible accumulation of errors in the profile and saves the need of time consuming incremental calculation. Sometimes we can go through to the terminal to simply obtain the final maximum load and elastic distortion of tooling. It may be also possible,starting from the final product,to continue an UBA analysis back to a desirable preform shape as was attempted by Kitahara et al. [92] who carried out this retrogression by FEM.

A more important and constructive part which UBA will take is the development of the intuition of engineers and investigators by cultivating better understanding of the physical phenomena, so that time and labour required for the design and attempt of new operations will be greatly reduced and even novel forming methods could be intuitively invented. The present author is convinced that the frequent and lighthearted interchange or dialogue between experimental observation with naked eyes and mathematical modelling by easy UBA does cultivate the understanding.

<u>List of references</u>

[1] Gvozdev, A.A., see I.J.M.S. 1 (1960), p. 322

[2] Hill, R.: On the State of Stress in a Plastic-Rigid Body at
     the Yield Point, Phil. Mag. 42 (1951), p. 868

[3] Green, A.P.: A Theoretical Investigation of the Compres-
     sion of a Ductile Material between Smooth Flat Dies, ibid.,
     p. 900

[4] Green, A.P.: Calculations on the Theory of Sheet Drawing,
     BISRA Report MW/B/7/52 (1952)

[5] Shield, R.T.; Drucker, D.C.: The Application of Limit Ana-
     lysis to Punch-Indentation Problems, Tr. A.S.M.E., J. Appl.
     Mech. 20 (1953), p. 453

[6] Shield, R.T.: The Plastic Indentation of a Layer by a Flat
     Punch, Quart. J. Appl. Math. 13 (1955), p. 27

[7] Levin, E.: Indentation Pressures of a Smooth Circular
     Punch, ibid., p. 133

[8] Hill, R.: New Horizons in the Mechanics of Solids, J. Mech.
     Phys. Solids 5 (1956), p. 66

[9] Kudo, H.: An Analysis of Plastic Compressive Deformation of
     Lamella between Rough Plates by Energy Method, Proc. 5. Ja-
     pan Nat. Congr. Appl. Mech. (1955), p. 75

[10] Hill, R.: The Mathematical Theory of Plasticity, Oxford
     Clarendon Press (1950)

[11] Kudo, H.: A Computation of Required Pressure for Extrusion-
     Forging Circular Shells, Proc. 7. Japan Jat. Congr. Appl.
     Mech. (1957), p. 57

[12] Kudo, H.: An Upper Bound Approach to Plane-Strain Forging
     and Extrusion — I & II, Report Aeron. Res. Inst., Tokyo Univ.
     1 (1958), p. 37 [J]; I.J.M.S. 1 (1960), p. 57; p. 229

[13] Johnson, W.: Over-Estimates of Load for Some Two-Dimension-
     al Forging Operations, Proc. 3. U.S. Nat. Congr. Appl. Mech.
     (1958), p. 571

[14] Johnson, W.: Estimation of Upper Bound Loads for Extrusion
     and Coining Operations, Proc. I.M.E. 173 (1959), p. 61

[15] Johnson, W.: An Elementary Consideration of Some Extrusion
     Defects, Appl. Sci. Res., Sect. A 8 (1959), p. 52

[16] Johnson, W.: Cavity Formation and Enfolding Defects in
     Plane-Strain Extrusions Using a Shaped Punch, ibid., p. 228

[17] Kudo, H.:   Some Analytical and Experimental Studies of Axi-
     symmetric Cold Forging and Extrusion—I, Report Aeron. Res.
     Inst.. Tokyo Univ. 1 (1958), p. 212 [J]; I.J.M.S. 2 (1960),
     p. 102

[18] Kudo, H.:   An Upper Bound Approach to a Simple Axisymmetric
     Closed-Die Forging, Proc. 10. Japan Nat. Congr. Appl. Mech.
     (1960), p. 145

[19] Kudo, H.:   An Upper-Bound Approach to Plane-Strain Forging
     and Extrusion—III, Report Aeron. Res. Inst., Tokyo Univ.,
     1 (1958), p. 131 [J]; I.J.M.S. 1 (1960), p. 366

[20] Kudo, H.:   Some Analytical and Experimental Studies of Axi-
     symmetric Cold Forging and Extrusion—II, Report Aeron. Res.
     Inst., Tokyo Univ. 1 (1959), p. 247 J ; I.J.M.S. 3 (1961),
     p. 91

[21] Tanner, R.I.; Johnson, W.:   Temperature Distributions in
     Some Fast Metal-Working Operations, I.J.M.S. 1 (1960), p. 28

[22] Johnson, W.; Kudo, H.:   The Use of Upper Bound Solutions
     for the Determination of Temperature Distributions in Fast
     Hot Rolling and Axisymmetric Extrusion Processes, ibid.,
     p. 175

[23] Tarnovskii, I. Ya.; Pozdeev, A.A.; Ganago, O.A.:   Deforma-
     tsii i Usiliya pri Obrabotke Metallov Davleniem, Mashinostr.
     Literatury, Moskva (1959) [R]

[24] Avitzur, B.; Yang, C.T.:   Analysis of Power Spinning of
     Cones, Tr. A.S.M.E., Ser. B 82 (1960), p. 231

[25] Yamada, Y.:   Yield-Point Load of Rigid-Plastic Body—I,
     Kikai-no-Kenkyu, Tokyo 10 (1958), p. 621 [J]

[26] Shindo, A.:   General Consideration on the Compression of a
     Wedge by Rigid Flat Die—I, Tr. J.S.M.E. 27 (1961), p. 447
     [J]; Bull. J.S.M.E. 5 (1962), p. 21

[27] Kudo, H. ; Takahashi, H.:   On Some Complete Solutions for
     Steady State Extrusion in Plane Strain, J. J.S.T.P. 5(1964)
     p. 237; p. 464 [J]

[28] Stepanskii, L.G.:   O Granitsakh Ochaga Plastitseskoi Defor-
     matsii pri Vydavlivanii, Best. Mashinostr., 43 (1963), p.
     59 [R]

[29] Avitzur, B.:   Analysis of Wire Drawing and Extrusion Coni-
     cal Dies of Small Cone Angle, Tr. A.S.M.E. Ser. 85 (1963),

p. 89

[30] Avitzur, B.:   Analysis of Metal Extrusion, Tr. A.S.M.E.,
     Ser. B 87 (1965), p. 57

[31] Halling, J.; Mitchell, L.A.:   An Upper Bound Solution for
     Axisymmetric Extrusion, I.J.M.S. 7 (1965), p. 277

[32] Adie, J.F.; Alexander, J.M.:   A Graphical Method of Obtain-
     ing Hodographs for Upper-Bound Solutions to Axisymmetric
     Problems, I.J.M.S. 9 (1967), p. 349

[33] Lambert, E.R.; Kobayashi, S.:   Admissible Velocity Fields
     for some Steady-State Forming Processes in Plane-Strain and
     Axisymmetry, Proc. J.S.M.E. Semi Intern. Symp., Tokyo
     (1967), p. 53

[34] Lambert, E.R.; Mehta, H.S.; Kobayashi, S.:   A New Upper-
     Bound Method for Analysis of Some Steady-State Plastic De-
     formation Processes, Tr. A.S.M.E., Ser. B 91 (1969), p. 731

[35] Nagpal, V.; Altan, T.:   Analysis of the Three-Dimensional
     Metal Flow in Extrusion of Shapes with the Use of Dual
     Stream Functions, Proc. 3. N.A.M.R.C. (1975), p. 26

[36] Nagpal, V: On the Solution of Three-Dimensional Metal-
     Forming Processes, Tr. A.S.M.E., Ser. B 99 (1977), p. 624

[37] Halling, J.; Mitchell, L.A.:   Use of Upper Bound Solutions
     for Predicting the Pressure for the Plane Strain Extrusion
     of Materials, J. Mech. Eng. Sci. 6 (1964), p. 240

[38] Haddow, J.B.:   Comment on "An Upper Bound Solution for Axi-
     symmetric Extrusion", I.J.M.S. 8 (1966), p. 145

[39] Thomason, P.F.:   A Theory for Ductile Fracture by Internal
     Necking of Cavities, J. Inst. Metals, 96 (1968), p. 360

[40] Zimmernan, Z.; Avitzur, B.:   Analysis of the Effect of
     Strain Hardening on Central Bursting Defects in Drawing and
     Extrusion, Tr. A.S.M.E., Ser. B 92 (1970), p. 135

[41] Avitzur, B.:   Strain-Hardening and Strain-Rate Effects in
     Plastic Flow through Conical Converging Dies, Tr. A.S.M.E.,
     Ser. B 89 (1967), p. 556

[42] Cristescu, N.:   Plastic Flow through Conical Converging
     Dies, Using a Viscoplastic Constitutive Equation, I.J.M.S.
     17 (1975), p. 425

[43] Lahoti, G.D.; Altan, T.:   Prediction of Temperature Distri-
     butions in Axisymmetric Compression and Torsion, Tr. A.S.M.

E., Ser. H 97 (1975), p. 113

[44] Fenton, R.G.: Effects of Ram Speed and Size on the Required Extrusion Pressure, Proc. 3. N.A.M.R.C. (1975), p. 41

[45] Avitzur, B.; Bishop, E.D.; Hahn, W.C.: Impact Extrusion-Upper Bound Analysis of the Early Stage, Tr. A.S.M.E., Ser. B 94 (1972), p. 1079

[46] Tirosh, J.; Kobayashi, S.: Kinetic and Dynamic Effects on the Upper Bound Loads in Metal-Forming Processes, Tr. A.S.M.E., Ser. E 98 (1976), P. 314

[47] Johnson, W.; de Malherbe, M.C.; Venter, R.: Upper Bounds to the Load for the Plane Strain Working of Anisotropic Metals, J. Mech. Eng. Sci. 14 (1972), p. 297

[48] Oyane, M.; Tabata, T.: Slip-Line Field Theory and Upper-Bound Theory for Porous Materials, J. J.S.T.P., 15 (1974), p. 43 [J]

[49] Avitzur, B.: The Production of Bi-Metal Wire, Wire J., 3 (1970), p. 42

[50] Osakada, K.; Limb, M.; Mellor, P.B.: Hydrostatic Extrusion of Composite Rods with Hard Cores, I.J.M.S. 15 (1973), p. 291

[51] Avitzur, B.; Grossman, G.: Hydrodynamic Lubrication in Rolling of Thin Strips, Tr. A.S.M.E., Ser. B 94 (1972), p. 317

[52] Lee, C.H.; Altan, T.: Influence of Flow Stress and Friction upon Metal Flow in Upset Forging of Rings and Cylinders, Tr. A.S.M.E. , Ser. B 94 (1972), p. 775

[53] Jain, S.C.; Bramley, A.N.; Lee C.H.; Kobayashi, S.: Theory and Experiment in Extrusion Forging, Proc. 11. Int.M.T.D.R. Conf. (1970), p. 1097

[54] Takahashi, H.; Murakami, T.: Effects of Tool Angle and Friction in an Open-Die Forging in Axisymmetry—I & II, J. J.S.T.P., 12 (1971), p. 31; p. 122 [J]

[55] Kasuga, Y.; Tsutsumi, S.; Saiki, H.: Research on Material Flow in Sunken Forging Dies—I & II, Tr. J.S.M.E. 39 (1973), p. 1353; p. 1366 [J]; Bull. J.S.M.E. 16 (1973), p. 1960

[56] McDermott, R.P.; Bramley, A.N.: An Elemental Upper Bound Technique for General Use in Forging Analysis, Proc. 15.

Int. M.T.D.R. Conf. (1974), p. 437

[57] Geiger, R.: Der Stofffluss beim kombinierten Napffliess-
pressen, Bericht Inst. Umformtechnik, Univ. Stuttgart, 36
(1976) [G]

[58] Johnson, W.; Kudo, H.: Plane-Strain Deep Indentation,
Proc. 5. Int. M.T.D.R. Conf. (1964), p. 441

[59] Nagpal, V.; Lahoti, G.D.; Altan, T.: A Numerical Method
for Simultanuous Prediction of Metal Flow and Temperatures
in Upset Forging of Rings, Tr. . A.S.M.E., Ser. B 100
(1978), p. 413

[60] Kudo, H.; Tamura, K.: Analysis and Experiment in V-Groove
Forming—I, II, III & IV, J. Jap. Soc. Precision Eng., 34
(1968), p. 38; 36 (1970), p. 256; p. 318; 37 (1971), p. 534
[J]; Ann. C.I.R.P., 17 (1969), p. 297

[61] Usui, E.; Masuko, M.: Fundamental Study on Three Dimensional Machining—I & II, Tr. J.S.M.E. 38 (1972), p. 3255;
p. 3264 [J]; Bull. J.S.M.E. 16 (1973), p. 1214

[62] Hayama, M.: On the Mechanism of Shear Spinning, Proc. 1.
I.C.P.E., Tokyo (1974), p. 262

[63] Oh, S.I.; Kobayashi, S.: An Approximate Method for a
Three-Dimensional Analysis of Rolling, I.J.M.S. 17 (1975),
p. 293

[64] Chen, C.T.; Ling, F.F.: Upper-Bound Solutions to Axisymmetric Extrusion Problems, I.J.M.S. 10 (1968), p. 863

[65] Nagpal, V.: General Kinematically Admissible Velocity
Fields for Some Axisymmetric Metal Forming Problems, Tr. A.
S.M.E., Ser. B 96 (1974), p. 1197

[66] Nagpal, V.: Analysis of Plane-Strain Extrusion through
Arbitrarily Shaped Dies Using Flow Function, Tr. A.S.M.E.,
Ser. B 99 (1977), p. 754

[67] Chang, K.T.; Choi, J.C.: Upper Bound Solutions to Symmetri-
cal Extrusion Problems through Curved Dies, Proc. 12. Mid-
western Conf. (1971), p. 383

[68] Busch, R.: Untersuchungen über das Abstreckziehen von
zylindrischen Hohlkörpern beim Raumtemperatur Bericht
Inst. Umformtech., Univ. Stuttgart 10 (1969) [G]; Indus-
trie Anzeiger 94 (1972), p. 609 [G]

[69] Andresen, K.: Blockstauchen zwischen ebenen parallelen

Bahnen, Arch. Eisenhütwes. 44 (1973), p. 595 [G]

[70] Lahoti, G.D.; Altan, T.: Prediction of Temperature Distributions in Tube Extrusion Using a Velocity Field without
Discontinuities, Proc. 2. N.A.M.R.C. (1974), p. 209

[71] Yang, D.-Y.; Lee, C.-H.: Analysis of Three-Dimensional Extrusion of Sections through Curved Dies by Conformal Transformation, I.J.M.S. 20 (1978), p. 541

[72] Yang, D.-Y.; Kim, M.-U.; Lee, C.-H.: An Analysis for Extrusion of Helical Shapes from Round Billets, ibid., p. 695

[73] Gunasekera, J.S.; Hoshino, S.: Analysis of Extrusion or
Drawing of Polygonal Sections through Straightly Converging Die, I.J.M.S. 24 (1982), p. 589

[74] Kiuchi, M.; Kishi, H.; Ishikawa, M.: Study on Non-Symmetric Extrusion and Drawing —I, J. J.S.T.P. 24 (1983), p.
290 [J]

[75] Kiuchi, M.; Ishikawa, M.: Study on Non-Symmetric Extrusion
and Drawing of Pipe, Seisan-Kenkyu, Res. Inst. Indust. Sci.,
Tokyo Univ. 33 (1981), p. 473

[76] Oudin, J.; Ravalard, Y.: A General Method for Computing
Plane Strain Plastic Flows, Proc. 20. Int. M.T.D.R. Conf.
(1979), p. 211

[77] Avitzur, B.; Iobst, J.W.; McDermott, R.P.: AXIFORM—A
Computer Simulation Program for Axisymmetric Forging and
Extrusion, Proc. 6. N.A.M.R.C. (1978), p. 174

[78] Osman, F.H.; Bramley, A.N.: Metal Flow Prediction in Forging and Extrusion Using UBET, Proc. 20. Int. M.T.D.R. Conf.
(1979), p. 51

[79] Thornton, J.N.; Bramley, A.N.: An Approximate Method for
Predicting Metal Flow in Forging and Extrusion Operations,
Proc. I.M.E. 194 (1980), p. 9

[80] Kiuchi, M.; Shigeta, S.: Application of UBET to Asymmetric
Forging Process, J. J.S.T.P. 22 (1981), p. 1208 [J]

[81] Gatto, F.; Giarda, A.: The Characteristics of the Three-
Dimensional Analysis of Plastic Deformation According to
Spacial Elementary Rigid Regions Method, I.J.M.S. 23 (1981),
p. 129

[82] Tomita, Y.; Seguchi, Y.; Shindo, A.; Tanaka, K.: Note on a
Modification of the Stream Function Method and its Applica-

tion to the Analysis of Metal Forming Processes, Proc. 8.
N.A.M.R.C. (1980), p. 144

[83] Kitahara, Y.; Osakada, K.; Fujii, S. Narutaki, R.:   Analysis of Plane-Strain Metal Forming Problem with Linear Programming Method, Bull. J.S.M.E., 22 (1979), p. 763

[84] Kudo, H.; Avitzur, B.; Yoshikai, T.; Luksza, J.; Moriyasu,
M.; Ito, S.:   Cold Forging of Hollow Cylindrical Components
Having an Intermediate Flange—UBET Analysis and Experiment,
Ann. C.I.R.P. 29 (1980), p. 129

[85] Kiuchi, M.; Hsiang, S.-H.:   Study on Application of Limit
Analysis to Rolling Process—I, J. J.S.T.P. 22 (1981), p.
927 [J]

[86] Ohga, K.; Kondo, K.; Jitsunari, T.:   Research on Precision
Die Forging Utilizing Divided Flow—I, II, III, IV, Tr. J.S.
M. E., Ser. C 48 (1982), p. 425; p. 435; p. 443; p. 436 [J]

[87] Price, J.W.H.; Alexander, J.M.:   A Study of the Isothermal
Forming or Creep Forming of a Titanium Alloy, Proc. 4. N.A.
M.R.C. (1976), p. 46

[88] Kudo, H.; Shinozaki, K.:   Investigation into Multiaxial Extrusion Process to Form Branched Parts, Proc. 1. I.C.P.E.,
Tokyo (1974), p. 314

[89] Johnson, W.; Mamalis, A.G.:   Force Polygons to Determine
Upper Bounds and Force Distribution in Plane Strain Metal
Forming Processes, Proc. 18. Int. M.T.D.R. Conf. (1977),
p. 11

[90] Kiuchi, M.; Murata, Y.:   Study on Application of UBET, Proc.
4. I.C.P.E., Tokyo (1980), p. 66

[91] Austen, A.R.; Avitzur, B.:   Influence of Hydrostatic Pressure on Void Formation at Hard Particles, Trans. A.S.M.E.,
Ser. B 96 (1974), p. 1192

[92] Kitahara, Y.; Osakada, K.; Fujii, S.; Narutaki, R.:   Analysis of Deformation of Plates in Free Forging Using Rigid-
Plastic FEM, J. J.S.T.P. 18 (1977), p. 753 [J]

Abbreviations:
    Proc. I.C.P.E. - Proceedings of International Conference
                     on Production Engineering
    J. J.S.T.P. - Journal of Japan Society for Technology of
                     Plasticity
    [G], [J], [R] - written in German, Japanese and Russian
                     resp.

Entwicklung von Stoffgesetzen für die Hochtemperaturplastizität

E.Steck, Institut für Allgemeine Mechanik und Festigkeitslehre,
Technische Universität Braunschweig/Germany

Summary

High temperature plasticity of metals at temperatures $> 0.5*$ melting
temperature is, at least for stress ranges of technical interest, explained by thermally activated dislocation movements, where the "structure" of the material, i.e. distribution and strength of the internal
barriers, which act against these movements, are of strong influence on
these processes.

The characterization of this structure by transition probabilities of a
discrete Markov-chain results in a stochastic model which is able to
represent essential and typical features which are characteristic for
high temperature plasticity, such as stationary creep, dependence of the
internal structure on stress and temperature or transition times to stationary creep after changes of these loads, in an at least qualitatively
satisfactory manner, and which allows for an examination of the effects
of assumptions about the deformation mechanicsms on the microscale.

0 Einleitung

Mit dem Begriff Hochtemperaturplastizität bezeichnet man das Verhalten
von kristallinen Festkörpern (Metalle, Fels, Steinsalz) bei Temperaturen
oberhalb etwa der halben Schmelztemperatur. Bei diesen hohen Temperaturen zeigt der Zusammenhang zwischen Spannungen und Verformungen eine
ausgeprägte Zeitabhängigkeit. Es existiert zwar eine von der Temperatur
abhängige Fließgrenze, bei deren Überschreiten spontane plastische Verformungen auftreten, im Gegensatz zum Verhalten bei niedrigen Temperaturen treten aber auch schon bei Spannungen unterhalb der Fließgrenze
ständig wachsende inelastische Formänderungen auf. Diese Erscheinung ist
sowohl für Umformvorgänge als auch für eine Reihe anderer technischer
Anwendungen von erheblichem Interesse, wie z.B. im Apparatebau, bei
Flugtriebwerken oder Energieerzeugungsanlagen, wo metallische Werkstoffe
in zunehmendem Maße bei hohen Temperaturen eingesetzt werden müssen. (In

modernen Hochtemperaturreaktoren werden z.B. Bauteile aus Stahlwerkstof-
fen Temperaturen von 900°C und darüber ausgesetzt.)

Die für die Hochtemperaturplastizität kennzeichnenden zeitabhängigen
Vorgänge (Kriechen, Relaxation) und ihre Wechselwirkung mit der spontanen Plastizität, sind der Rechnung bisher nur unvollkommen zugänglich.
Die Stoffgesetze, die den Berechnungen zugrundegelegt werden können,
sind bisher im wesentlichen aus dem Experiment abgeleitete Beziehungen
für das stationäre (sekundäre) Kriechen, mit einer gewissen Unterstützung durch Vorstellungen über die Ursachen dieser Verformungen, die sie
mit den selben Vorgängen in den Materialien verknüpfen, die auch für die
plastischen Verformungen bei niedrigeren Temperaturen verantwortlich
sind. Eine theoretische Beschreibung der Vorgänge, die bei Last- oder
Temperaturwechseln auftreten, steht noch aus. Die Kriechdaten (z.B.
Kriechkonstante und Kriechexponent im Nortonschen Kriechgesetz) müssen
für die eingesetzten Werkstoffe in Langzeitversuchen ermittelt werden.
Dies gilt auch für die in der technischen Praxis sehr wichtigen Daten
für die Zeitstandsfestigkeit in Abhängigkeit von Temperatur und Spannung.

Es ist weitgehend akzeptiert, daß die Mechanismen, die zu den zeitabhängigen Deformationen bei hohen Temperaturen führen, im wesentlichen die
selben sind, wie diejenigen, die das spontane plastische Verhalten
verursachen, nämlich Abgleitvorgänge in den Kristalliten, die durch
Versetzungen unterstützt werden. Diese Auffassung wird durch die Beobachtung unterstützt, daß die Kriechvorgänge im wesentlichen inkompres-
sibel ablaufen, und daß bei mehrachsiger Beanspruchung die Invarianten
des Spannungsdeviators (hauptsächlich die zweite Invariante) die selbe
maßgebende Rolle als die Beanspruchung kennzeichnende Größen spielen,
wie bei spontaner Plastizität.

Wegen der großen technischen Bedeutung der Hochtemperaturplastizität und
der interessanten werkstoffphysikalischen Phänomene, die mit ihr verbunden sind, ist sie schon seit langer Zeit Gegenstand vielfältiger For-
schungsaktivitäten. Diese Arbeiten haben aber bisher nicht zu einer
befriedigenden, physikalisch begründeten, kontinuumsmechanischen Theorie
geführt, die Stoffgesetze zur Verfügung stellt, die in numerischen

Berechnungen des Verhaltens von Bauteilen bei hohen Temperaturen ange-
wendet werden könnten. Ein wesentlicher Grund hierfür dürfte darin zu
suchen sein, daß der Versuch, dieses Verhalten mathematisch zu beschreiben, entweder von einem rein phänomenologischen Standpunkt oder von
einem rein metallphysikalischen Standpunkt aus vorgenommen wurde, und
dabei jeweils wesentliche Gesichtspunkte der anderen Anschauungsweise
außer Betracht blieben.

1 Phänomenologische Stoffgesetze

Die phänomenologischen, kontinuumsmechanische Theorien, die in neuerer
Zeit auf diesem Gebiet aufgestellt wurden, sind im wesentlichen aus
thermodynamischen Rahmentheorien abgeleitet, bei denen das Werkstoffverhalten durch innere Variablen erfasst werden soll, deren Entwicklungsgleichungen die Änderungen der Werkstoffstruktur beschreiben. Die Rah-
mentheorie hat die Zulässigkeit der Ansätze zu garantieren. Die metallphysikalischen Untersuchungen sind naturgemäß stark auf eine detaillier-
te Beschreibung der Vorgänge im Mikrobereich gerichtet. Die Übertragung
der Ergebnisse auf die makroskopische Ebene geschieht durch - oft sehr
stark vereinfachende - Mittelungsprozesse.

Die in jüngerer Zeit in der Literatur vorgeschlagenen phänomenologischen, kontinuumsmechanischen Modelle für die Hochtemperaturplastizität
von Metallen gehen häufig von Ansätzen aus, die die inelastischen Formänderungsgeschwindigkeiten in der Form

$$\dot{\underline{\varepsilon}}_{ie} = f(\underline{\underline{\sigma}}, T, S_i) \tag{1.1}$$

als Funktion der äußeren Spannung, der Temperatur und der Werkstoff-
"Struktur" beschreiben, wobei die eigentliche Beschreibung des Werkstoffverhaltens durch Beziehungen für die zeitliche Änderung der Struk-
turparameter S geschieht, von denen man z.B. annimmt, daß sie in der
Form

$$\dot{S}_i = g_i^v(\dot{\underline{\varepsilon}}_{ie}, T, S_i) - g^\varepsilon(T, S_i, \underline{\underline{\sigma}}) \tag{1.2}$$

als von Verfestigungs- und Erholungsvorgängen abhängig angenommen werden
können.

Die Größen S_i , über die die Strukturänderungen beschrieben werden, sind dabei entweder makroskopische Verfestigungsparameter, mit denen isotrope und kinematische Verfestigungsanteile in die Stoffgesetze aufgenommen werden [1], oder - bei thermodynamisch begründeten Ansätzen - innere Variable, mit denen man nicht direkt von außen beeinflußbare, mikroskopische Vorgänge, die das Werkstoffverhalten bestimmen, zu beschreiben sucht [2-5].

Da in beiden Fällen die Begründung der speziellen Wahl der Parameter, sowie ihrer Stellung in den Stoffgleichungen und die Form der für die Beschreibung ihrer Veränderung während der betrachteten Vorgänge erforderlichen Entwicklungsgleichungen bis zu einem gewissen Grade willkürlich bleibt, muß nach einem Weg gesucht werden, die erforderlichen Parameter und ihre Entwicklungsgleichungen aus Kenntnissen zu gewinnen, die von der Metallphysik über die Mechanismen der Hochtemperaturplastizität zur Verfügung gestellt werden.

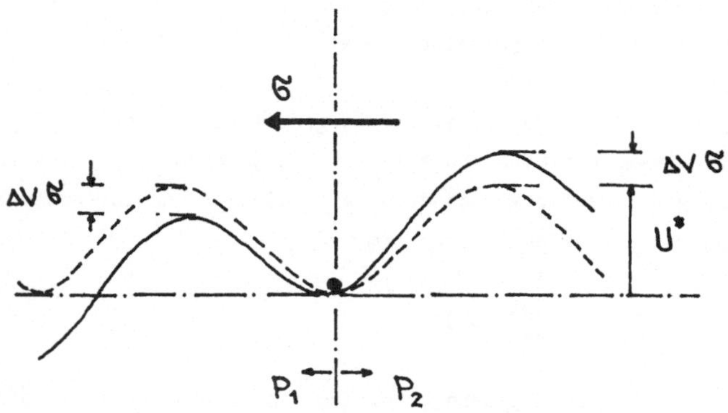

Bild 1: Thermisch aktivierte Platzwechsel

Hierzu wurde ein kinetisches Modell aufgestellt und untersucht, das die Hochtemperaturplastizität über thermisch aktivierte Platzwechsel beschreibt, und das auf die Behandlung des Kriechens und der Plastizität über einen M a r k o v -Prozess führt. Die erhaltene M a r k o v - K e t t e zeigt eine Reihe von Eigenschaften, die für Kriechvorgänge charakteristisch sind, z.B. die Existenz stationärer Zustände bei konstanter äußerer Spannung und Temperatur oder die typische Zeitabhängigkeit der Größe der inelastischen Formänderungsgeschwindigkeiten bei Spannungs- und Temperaturwechseln.

2 Stochastische (kinetische) Modelle

Besonders in den vierziger Jahren wurden Untersuchungen durchgeführt, bei denen der Versuch gemacht wurde, die Kriechvorgänge aus der Annahme von thermisch aktivierten Versetzungsbewegungen zu erklären. Bild 1 zeigt die zugrundeliegende Vorstellung. Die Verformungsmechanismen (Versetzungen oder Versetzungspakete) liegen im Kristallgitter vor Hindernissen, die durch eine Potentialschwelle der Höhe U^* gekennzeichnet sind. Die angelegte Spannung σ vermindert die Hindernishöhe in ihrer Wirkrichtung um den Betrag $\Delta V \, \sigma$ (ΔV ist das sog. Aktivierungsvolumen), und erhöht es in Gegenrichtung um den selben Wert. Die Annahme, daß die Energie der Fließeinheiten einer Boltzmann-Verteilung folgt, führt zu Wahrscheinlichkeiten für die Überwindung der Hindernisse durch die Fließeinheiten in Spannungsrichtung die proportional zu

$$p_1 = A \exp\left(- \frac{U^* - \Delta V \sigma}{k_B T}\right) \qquad (2.1)$$

sind. (k_B ist die Boltzmannkonstante, T die absolute Temperatur.) In der Gegenrichtung ist die Wahrscheinlichkeit, das Hindernis zu überwinden durch

$$p_2 = A \exp\left(- \frac{U^* + \Delta V \sigma}{k_B T}\right) \qquad (2.2)$$

gegeben. Mit der Bewegung der Verformungsmechanismen im Gitter ist eine makroskopische inelastische Formänderungsgeschwindigkeit verbunden, die proportional zur Differenz p1-p2 angenommen wird. Das Modell führt damit für die inelastischen Formänderungsgeschwindigkeiten auf einen Ausdruck der Form

$$\dot{\varepsilon}_{ie} \sim A \exp\left(- \frac{U^*}{k_B T}\right) \sinh\left(\frac{\Delta V \sigma}{k_B T}\right) \qquad (2.3)$$

Dieser Ansatz war bei der qualitativen Erklärung experimenteller Befunde des stationären Kriechverhaltens metallischer Werkstoffe durchaus erfolgreich. Es ergaben sich Beziehungen der Form

$$\dot{\varepsilon}_{ie} = A(T) \; \sinh(B(T) \, \sigma) \tag{2.4}$$

die für hohe Spannungen in die Beziehung

$$\dot{\varepsilon}_{ie} = \bar{A}(T) \; \exp(B(T) \, \sigma) \tag{2.5}$$

und für niedrige Spannungen in

$$\dot{\varepsilon}_{ie} = \tilde{A}(T) \, \sigma \tag{2.6}$$

übergeführt werden können. Die durch den Faktor $\exp(-U/(k_B T))$ gegebene Temperaturabhängigkeit der Kriechvorgänge ist durch das Experiment gut bestätigt. Dies deutet darauf hin, daß die thermische Aktivierung von Verformungsmechanismen bei der Hochtemperaturplastizität tatsächlich eine wesentliche Rolle spielt. Es gelang jedoch auf der Basis dieses Modells nicht, die experimentell gut abgesichterte Beziehung (Nortonsches Kriechgesetz):

$$\dot{\varepsilon}_{ie} = A(T) \, \sigma^n \tag{2.7}$$

zu erklären. Man half sich mit einem nicht weiter begründeten Ansatz, bei dem in die vorhandenen Beziehungen ein Kriechexponent eingeführt wird, der experimentell bestimmt werden muß.

Die Beschränkung der Betrachtung auf eine einzige Hindernishöhe führte dazu, daß mit derartigen Modellen nur stationäres Verhalten behandelt werden konnte, da die Veränderung der Werkstoffstruktur nicht beschrieben wird. Täubert [6] schlug im Jahre 1958 eine Erweiterung dieser Vorstellungen vor, bei der die Veränderung der Werkstoffstruktur durch Erholungs- und Verfestigungsvorgänge mitberücksichtigt wird.

3 Erweitertes kinetisches Modell

Die Erweiterung besteht darin, daß nicht von einer einzigen, mit der Aktivierungsenergie der Selbstdiffusion in Verbindung gebrachten Hindernishöhe für die Platzwechsel der Gleitmechanismen ("Fließeinheiten",

flow-units) ausgegangen wird, sondern für diese Hindernisse ein Phasen-
raum vorgegeben wird, über den die Fließeinheiten, deren Zahl konstant
sein soll, je nach der Vorgeschichte, die der Werkstoff erfahren hat,
verteilt sind (Bild 2). Außerdem wird ein Erholungsmechanismus eingeführt, bei dem die Vorgabe gemacht wird, daß die Erholung um so stärke-
ren Einfluß auf die Veränderung der Werkstoffstruktur besitzt, je weiter
die Verfestigung fortgeschritten ist.

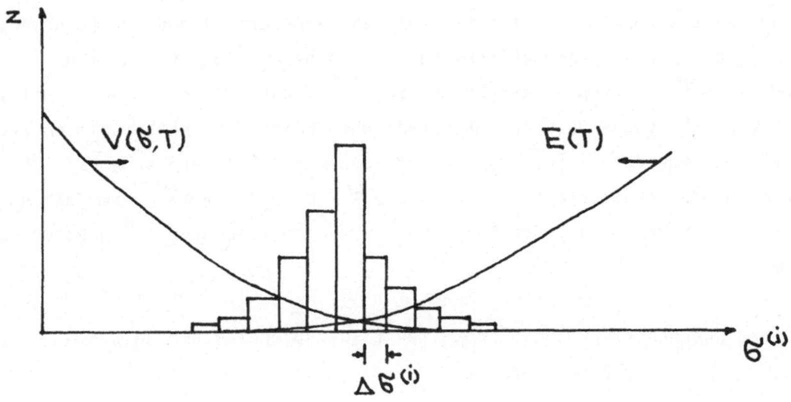

Bild 2: Verteilung der Fließeinheiten über der Zustandsachse

Die Beschäftigung mit diesem Ansatz zeigte, daß die Annahme einer Verteilung der Gleitmechanismen über ein Spektrum von Hindernisstärken,
sowie die Beschreibung der Verfestigung durch Annahme eines Platzwech-
sels zu einem höheren Hindernis, und der Erholung durch einen thermisch
aktivierten Vorgang, der von der Höhe der Verfestigung abhängt und nicht
zu makroskopischer Verformung führt, grundlegende Mechanismen annimmt,
die in vielen Vorschlägen, sowohl von Seite der Kontinuumsmechanik als
auch der Werkstoffkunde für Modelle zur Behandlung des Kriechens wieder-
zufinden sind. Darüberhinaus zeigte es sich, daß die Täubertsche Grund-
gleichung unmittelbar mit der Beschreibung der Vorgänge als Markov-
Ketten verknüpft werden kann. Diese Tatsache stellt für die Untersuchun-

gen besonders leistungsfähige mathematische Hilfsmittel zur Verfügung [7-9].

Die Berücksichtigung von Verfestigungs- und Erholungsvorgängen führt zu folgender Formulierung (Bild 2):

Bei den für die inelastischen Formänderungen verantwortlichen Gleitvorgängen haben die Gleitmechanismen Hindernisse zu überwinden, die durch innere Spannungen verschiedener Höhe repräsentiert werden können. Gleitschritte, die zur makroskopischen Formänderung beitragen, führen dazu, daß Verfestigung auftritt. Dies wird im Rechenmodell dadurch berücksichtigt, daß der Gleitmechanismus für den nächsten Schritt ein um den Betrag $\Delta \sigma^{(i)}$ höheres Hindernis zu überwinden hat. Die thermische Aktivierung des Vorgangs wird in der klassischen Weise dadurch berücksichtigt, daß eine von der Temperatur und der angelegten Spannung abhängige Übergangswahrscheinlichkeit von der Lage $\sigma^{(i)}$ nach $\sigma^{(i)} + \Delta \sigma^{(i)}$ im Zustandsraum des Systems, der durch die inneren Gegenspannungen $\sigma^{(i)}$ gegeben ist, existiert.

Diese Übergangswahrscheinlichkeit wird entsprechend der in Gleichung (2.1) gegebenen Form zu

$$\mathcal{P}(\sigma^{(i)} \rightarrow \sigma^{(i)} + \Delta \sigma^{(i)}) = V_{i,i+1}$$

$$= C_1 \Delta t \left(e^{-\frac{U^* - \Delta V(\sigma - \sigma^{(i)})}{k_B T}} \underset{(-)}{+e} ^{-\frac{U^* + \Delta V(\sigma + \sigma^{(i)})}{k_B T}} \right)$$

$$\longrightarrow \quad V_{i,i+1} = 2C_1 \Delta t \, \cosh\left(\frac{\Delta V \sigma}{k_B T}\right) e^{-\frac{U^* + \Delta V \sigma^{(i)}}{k_B T}} \qquad (3.1a)$$

oder

$$V_{i,i+1} = 2C_1 \Delta t \, \sinh\left(\frac{\Delta V \sigma}{k_B T}\right) e^{-\frac{U^* + \Delta V \sigma^{(i)}}{k_B T}} \qquad (3.1b)$$

angesetzt. U^* ist die Aktivierungsenergie der Selbstdiffusion im betrachteten Werkstoff. Die durch sie gegebene Hindernishöhe wird durch die

äußere Spannung in deren Wirkrichtung um den Betrag $\Delta V \sigma$ vermindert
und durch die infolge der Verfestigung wachsenden inneren Gegenspannungen um den Betrag $\Delta V \sigma^{(i)}$ erhöht.

Weiter wird angenommen, daß mit den Platzwechseln von $\sigma^{(i)}$ nach $\sigma^{(i)} + \Delta \sigma^{(i)}$
eine makroskopische Formänderungsgeschwindigkeit verbunden ist, die sich
zu

$$\dot{\varepsilon}_{ie} = C_1 \lambda \sum_{i=1}^{\infty} z_i \exp\left(- \frac{U^* - \Delta V(\sigma - \sigma^{(i)})}{k_B T}\right) \tag{3.2}$$

ergibt.

Steht der "cosh" in der Gleichung (3.1), so tritt Verfestigung auch ohne
Verformung auf. Dies entspräche etwa der Berücksichtigung von Alterungs-
vorgängen. Für den Fall, daß die Gleichung "sinh" enthält, tritt Verfestigung des Werkstoffs nur in Verbindung mit Verformung auf. Beide Argu-
mentationen sind möglich. Zunächst soll jedoch davon ausgegangen werden,
daß $x = (\Delta V \sigma / (k_B T))$ so groß ist, daß in guter Näherung sowohl cosh(x)
als auch sinh(x) durch 0.5*exp(x) ersetzt werden kann.

Der Verfestigung wirkt ein Erholungsmechanismus entgegen, von dem ebenfalls angenommen wird, daß er thermisch aktiviert ist, wobei die trei-
bende Kraft, die die Fließeinheiten wieder zu niedrigeren Hindernishöhen
bewegt, durch die inneren Spannungen gegeben ist, die mit den durch die
Verfestigungsvorgänge erzeugten Gitterverspannungen verbunden sind. Auch
dieser Vorgang geschieht im Zeitintervall Δt mit einer bestimmten Übergangswahrscheinlichkeit die zu

$$P(\sigma^{(i)} \rightarrow \sigma^{(i)} - \Delta \sigma^{(i)}) = E_{i,i-1}$$
$$= C_2 \Delta t \, e^{- \frac{Q^* - \Delta W \sigma^{(i)}}{k_B T}} \tag{3.3}$$

angesetzt wird. Q^* ist die diesen Vorgang bestimmende Aktivierungsenergie, ΔW das zugehörige Aktivierungsvolumen.

Die Parameter Q^*, ΔV, ΔW, C_1, C_2 und λ müssen durch Vergleich mit den
Versuchsergebnissen quantitativ bestimmt werden. Für U^* wird im folgen-

den, in Übereinstimmung mit experimentellen Befunden [10], stets die
Aktivierungsenergie der Selbstdiffusion eingesetzt.

Eine gegebene Verteilung der Fließeinheiten ändert ihre Lage im Phasen-
raum mit Überganswahrscheinlichkeiten nach Gl.(3.1) zu höheren inneren
Spannungen und mit Übergangswahrscheinlichkeiten gemäß Gl.(3.3) zu niedrigeren Werten. Die Verbleibewahrscheinlichkeit ist damit durch

$$B_{i,i} = 1 - V_{i,i+1} - E_{i,i-1} \qquad (3.4)$$

gegeben.

Mit diesen Wahrscheinlichkeiten läßt sich die zeitliche Änderung der
wahrscheinlichen Zahl der Gleitmechanismen an einer beliebigen Stelle
der Zustandsachse durch die Beziehung

$$\underline{z}(t+\Delta t) = \underline{\underline{SM}}\,\underline{z}(t) \qquad (3.5)$$

beschreiben. Hierbei ist

$$\underline{z}(t) = Z\,\underline{p}(\sigma^{(i)}, t) \qquad (3.6)$$

der Vektor, der die Verteilung der Zahl der aktiven Gleitmechanismen
über der Größe $\sigma^{(i)}$ angibt. Der Vektor $\underline{p}(\sigma^{(i)}, t)$ gibt die Verteilung der
Aufenthaltswahrscheinlichkeiten zur Zeit t an, Z ist die Gesamtzahl der
Gleitmechanismen.

$\underline{\underline{SM}}$ enthält die gemäß den Gleichungen (3.1), (3.3) und (3.4) formulierten
Übergangs- und Bleibewahrscheinlichkeiten in der Form

$$\underline{\underline{SM}} = \begin{bmatrix}
\vdots & & \\
E_{i-1,i-2} & 0 & 0 \\
B_{i-1,i-1} & E_{i,i-1} & 0 \\
V_{i-1,i} & B_{i,i} & E_{i+1,i} \\
0 & V_{i,i+1} & B_{i+1,i+1} \\
0 & 0 & V_{i+1,i+2} \\
& & \vdots
\end{bmatrix} \qquad (3.7)$$

SM ist eine spaltensummenkonstante Matrix mit der Spaltensumme 1, es

handelt sich also um eine stochastische Matrix, und die durch (3.5)
gegebene Entwicklung der Zustände des Systems stellt eine Markov-Kette
dar.

Die Behandlung der Vorgänge mit einem Markovschen Modell setzt voraus,
daß das System so geartet ist, daß die Kenntnis seines augenblicklichen
Zustands ausreicht, um sein Verhalten im nächsten Zeitschritt vollstän-
dig zu beschreiben. Diese Eigenschaft, die eine Behandlung der kontinuumsmechanischen Aufgabe durch Differentialgleichungssysteme (im Gegensatz zur Beschreibung mit Zeitfunktionalen) zuläßt, wird bei der Behandlung von plastizitätstheoretischen Aufgaben stets vorausgesetzt und
durch das Experiment bestätigt. Da kein grundlegender Unterschied zwischen den Verformungsmechanismen der "spontanen" Plastizität und dem
plastischen Verhalten kristalliner Werkstoffe bei hohen Temperaturen
angenommen werden muß, kann sie auch für Kriech- und Relaxationsvorgänge
als gültig vorausgesetzt werden.

3.1 Mathematische Eigenschaften der Markov-Kette

Mit den durch die Gleichungen (3.1) und (3.3) definierten Übergangswahrscheinlichkeiten erhält man eine unzerlegbare, tridiagonale, stochastische Matrix, die diagonalähnlich ist, da die außerhalb der Diago-
nale stehenden Koeffizienten sämtlich verschieden von Null und positiv
sind. Die Matrix $\underline{SM}$ besitzt damit ausschließlich reelle Eigenwerte. Da
der Spektralradius von stochastischen Matrizen Eins ist, besitzt $\underline{SM}$
einen maximalen Eigenwert $\lambda_1 = 1$. Für alle weiteren Eigenwerte gilt $\lambda_i < 1$.

Die Matrix $\underline{SM}$ läßt sich damit durch eine Hauptachsentransformation auf
Diagonalgestalt überführen. Man erhält

$$\underset{\sim}{\underline{SM}} = \underline{M}^{-1}\,\underline{SM}\,\underline{M} = \begin{bmatrix} 1 & 0 & 0 & 0 & 0 \\ 0 & \lambda_2 & 0 & 0 & 0 \\ 0 & 0 & \lambda_3 & 0 & 0 \\ 0 & 0 & 0 & \lambda_4 & 0 \\ 0 & 0 & 0 & 0 & \lambda_i \end{bmatrix} \qquad (3.8)$$

Die Transformationsmatrizen sind die Modalmatrix $\underline{M}$ (Matrix der spalten-
weise angeordneten Eigenvektoren) und ihre Inverse. Wendet man diese
Transformation auf die Markov-Kette nach Gleichung (3.5) an, so gelten

die Beziehungen

$$\tilde{\underline{z}} = \underline{M}^{-1}\underline{z} \qquad \underline{z} = \underline{M}\,\tilde{\underline{z}}$$

$$\underline{z}(t+\Delta t) = \underline{SM}\,\underline{z}(t)$$

$$\tilde{\underline{z}}(t+\Delta t) = \underline{M}^{-1}\underline{SM}\,\underline{z}(t)$$

$$= \underline{M}^{-1}\underline{SM}\,\underline{M}\,\tilde{\underline{z}}(t)$$

$$\tilde{\underline{z}}(t+\ t) = \tilde{\underline{SM}}\,\tilde{\underline{z}}(t) \tag{3.9}$$

zur Beschreibung der Entwicklung der Verteilung.

Solange man Spannung und Temperatur konstant hält, ändern sich die Koeffizienten der stochastischen Matrix nicht. Der Prozess läuft __homogen__ ab. Für diesen Fall lassen sich die Übergangswahrscheinlichkeiten nach n Schritten durch wiederholtes Anwenden der Matrix $\underline{SM}$ auf den jeweiligen Zustandsvektor gemäß

$$\underline{z}(t_0+\Delta t) = \underline{SM}\,\underline{z}(t_0)$$

$$\underline{z}(t_0+2\Delta t) = \underline{SM}\,\underline{z}(t_0+\Delta t) = \underline{SM}^2 \underline{z}(t_0)$$

$$\underline{z}(t_0+n\Delta t) = \underline{SM}^n \underline{z}(t_0) \tag{3.10}$$

bestimmen. Man erhält also offensichtlich direkt den Übergang von t_0 auf t_0+ n Δt durch Erheben der Matrix $\underline{SM}$ in die n-te Potenz. Dasselbe gilt für die Beschreibung des Vorgangs im Hauptachsensystem. Man erhält

$$\tilde{\underline{z}}(t_0+n\Delta t) = \tilde{\underline{SM}}^n \tilde{\underline{z}}(t_0) = \begin{bmatrix} 1 & 0 & 0 \\ 0 & \lambda_2^n & 0 \\ 0 & 0 & \lambda_i^n \end{bmatrix} \tilde{\underline{z}}(t_0) \tag{3.11}$$

Diese Beziehung zeigt unmittelbar, daß der homogene Vorgang einem stationären Wert der Verteilung und damit auch der inelastischen Formänderungsgeschwindigkeit zustrebt. Da die Eigenwerte λ_i für $i \neq 1$ sämtlich kleiner als 1 sind, verschwindet ihr Einfluß mit wachsender Schrittzahl. Der Zustandsvektor nimmt, unabhängig von der Ausgangsverteilung, einen stationären Wert an, der von der herrschenden Spannung und Temperatur bestimmt wird. Diese stationäre Verteilung entspricht dem Eigenvektor zum größten Eigenwert $\lambda_1 = 1$. Damit lassen sich die stationäre Verteilung und die zugehörige Formänderungsgeschwindigkeit bei gegebener Spannung und Temperatur unmittelbar angeben.

Die Berechnung von Transienten wird zweckmäßigerweise so durchgeführt, daß die Beziehung (3.11) ausgenutzt werden kann. Solange Temperatur und Spannung unverändert sind (homogener Prozess), kann die Verteilung für einen Zeitpunkt $t_o + n\Delta t$ aus der Verteilung für den Zeitpunkt t_o über diese Gleichung berechnet werden. Dies ermöglicht eine beliebige Wahl des Zeitmaßstabes, in dem die Berechnungen durchgeführt werden.

3.2 Anpassung von Parametern

Zur quantitativen Bestimmung der Paramter des Modells für verschiedene Werkstoffe wurden zunächst Versuchsergebnisse für stationäres Kriechen herangezogen, da Meßergebnisse für instationäre Vorgänge nur in geringer Zahl zur Verfügung stehen.

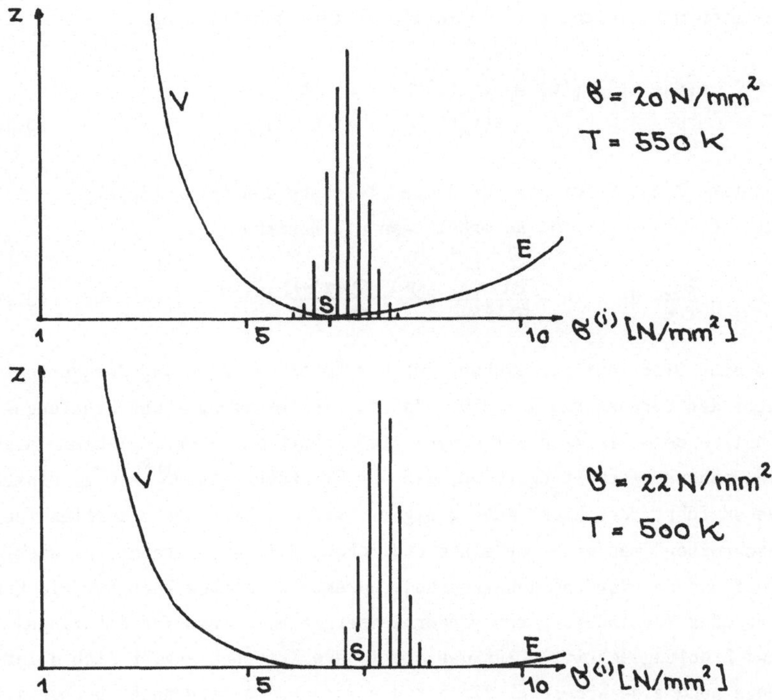

Bild 3: Stationäre Verteilungen für Reinstaluminium

Für die Anpassung wurde in einem ersten Schritt eine Näherung für die
stationäre Kriechgeschwindigkeit benutzt. Bild 3 zeigt stationäre Verteilungen für Reinstaluminium (die Werkstoffdaten wurden der Arbeit [1]
entnommen) bei verschiedenen Temperaturen und Spannungen. Die Ermittlung
der zugehörigen Parameter wird im folgenden dargestellt. Man erkennt,
daß sich die Verteilung der Fließeinheiten um den Schnitt der ebenfalls
eingetragenen Verfestigungs- und Erholungsfunktionen anordnet. Für diese
Lage der Verteilung halten sich die Verfestigungs- und Erholungsanteile
im Mittel die Waage. Gleichung (3.5) gibt als i-te Komponente des Vek-
tors $\underline{z}(t+\Delta t)$ den Ausdruck

$$z_i(t+\Delta t) = z_{i-1}(t)V_{i-1,i} + z_i(t)B_{i,i} + z_{i+1}(t)E_{i+1,i}, \tag{3.12}$$

der sich mit Gleichung (3.4) in die Differenzengleichung

$$z_i(t+\Delta t) - z_i(t) = -(z_i(t)V_{i,i+1} - z_{i-1}(t)V_{i-1,i}) $$
$$+ (z_{i+1}(t)E_{i+1,i} - z_i(t)E_{i,i-1}) \tag{3.13}$$

umformen läßt. Führt man für diese Gleichung den Grenzübergang
$\Delta t \to 0, \Delta \vartheta^{(i)} \to 0$ durch, so erhält man die Beziehung

$$\frac{\partial z}{\partial t} dt = -\frac{\partial (zV)}{\partial \vartheta^{(i)}} d\vartheta^{(i)} + \frac{\partial (zE)}{\partial \vartheta^{(i)}} d\vartheta^{(i)} \tag{3.14}$$

die eine Differentialgleichung für die zeitliche Änderung der Verteilung
unter der Voraussetzung ergibt, daß die Verfestigungs- und Erholungs-
schritte beim Platzwechsel einer Fließeinheit differentiell klein sind.
Eine nähere Betrachtung zeigt, daß der Grenzübergang $\Delta \vartheta^{(i)} \to 0$ zu einem
System führt, das nicht mehr geeignet ist, die Vorgänge zutreffend zu
beschreiben (man erwartet einen endlichen, für den betrachteten Werkstoff und den Verfestigungszustand charakteristischen Wert für die Erhöhung oder Verminderung der Hindernisenergie bei einem Verfestigungs-
bzw. Erholungsschritt). Für das stationäre Kriechen ergibt sich allerdings noch eine Näherung, die für die Parameteranpassung brauchbar ist.

Setzt man in Gleichung (3.14) die zeitliche Änderung der Verteilung zu

Null, so erhält man die Differentialgleichung

$$\frac{\partial}{\partial \sigma}(zE - zV) = 0 \qquad\qquad (3.15)$$

die unter der Bedingung, daß die Größe Z konstant sein muß, die Lösung

$$z = Z\,\delta(\sigma_s^{(i)} - \sigma^{(i)}) \qquad\qquad (3.16)$$

mit $\sigma_s^{(i)}$ aus $\qquad\qquad V(\sigma_s^{(i)}) = E(\sigma_s^{(i)})$

besitzt. Die Verteilung zieht sich damit am Schnittpunkt der Verfestigungs und Erholungsfunktion auf eine Deltafunktion mit dem Integral Z zusammen. Man erhält bei gegebener Spannung und Temperatur nur eine einzige Hindernishöhe, an der alle Gleitmechanismen liegen. Einsetzen dieser Lösung in die Beziehung (3.2) für die inelastischen Formänderungsgeschwindigkeiten führt auf das Ergebnis

$$\bar{\dot{\varepsilon}}_s = C_1 \lambda Z\, e^{-\dfrac{U^* - \Delta V(\sigma - \sigma_s^{(i)})}{k_B T}} \qquad\qquad (3.17)$$

das den Beziehungen entspricht, die von den älteren Kriechtheorien zur Verfügung gestellt werden, allerdings mit der zusätzlichen Eigenschaft, daß die Lage der Verteilung durch die Voraussetzungen über die Verfestigungs- und Erholungsmechanismen bestimmt ist.

Ein Vergleich stationärer Formänderungsgeschwindigkeiten, die einmal mit Gleichung (3.17) und zum andern für stationäre Verteilungen aus der Markov-Kette berechnet wurden, zeigen, daß die stationäre Kriechgeschwindigkeit, die mit Gleichung (3.17) berechnet wird, eine gute Näherung für die Ergebnisse aus (3.2) darstellt. Sie kann damit zur Parameteranpassung herangezogen werden.

Der Schnitt der Erholungs- und der Verfestigungsfunktion, und damit die Lage der "Nadel"-Verteilung ergibt sich mit $\Delta V = \Delta V_0 T$ und $\Delta W = \Delta W_0 T$, also der Annahme einer linearen Temperaturabhängigkeit der Aktivierungsvolumina, zu

$$\sigma_s^{(i)} = \frac{1}{1 + \dfrac{\Delta W_0}{\Delta V_0}}\left(\ln\left(\frac{C_1}{C_2}\right)\frac{k_B}{\Delta V_0} + \frac{1}{T\Delta V_0}(Q^* - U^*) + \sigma\right) \qquad (3.18)$$

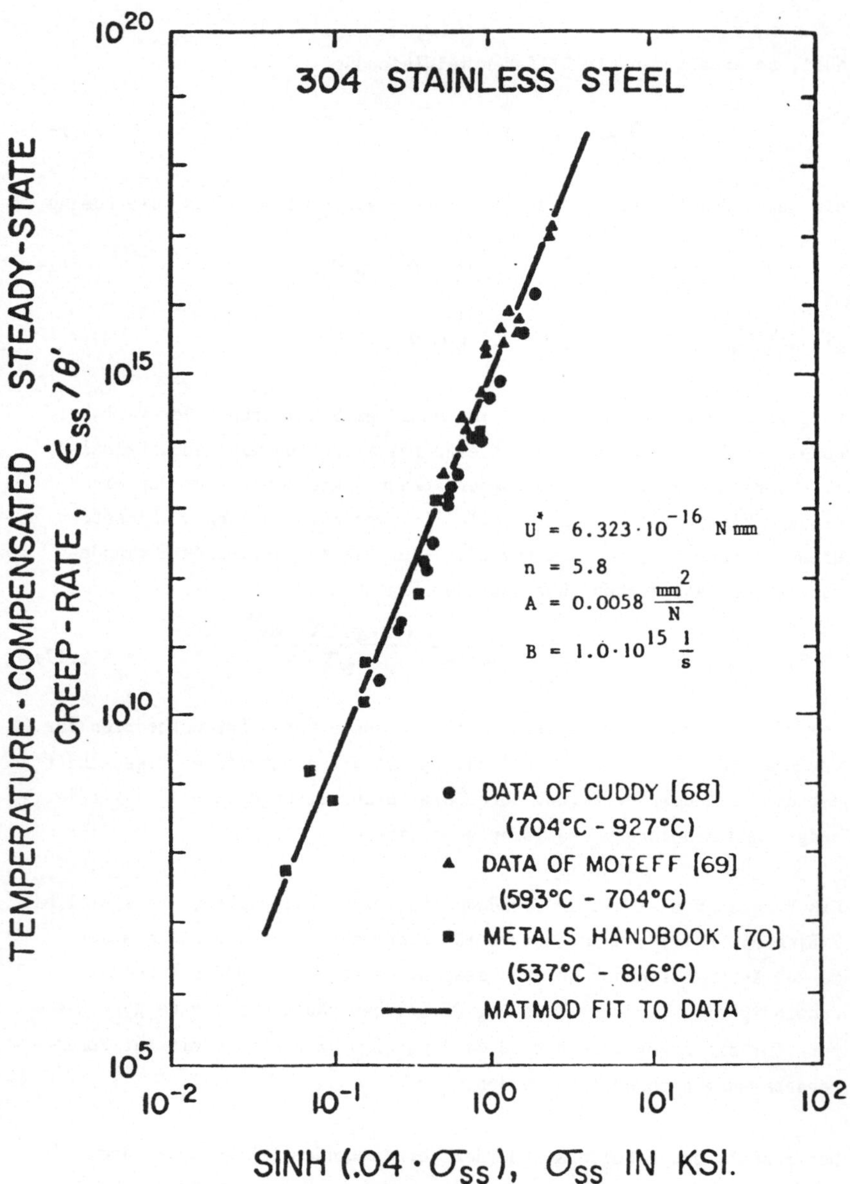

Bild 4: Stationäre Kriechdaten für 304 SS nach Miller [1]

Einsetzen dieser Größe in die Beziehung (3.17) liefert die im folgenden benutzte Näherung für die stationäre Kriechgeschwindigkeit in Abhängigkeit von Spannung und Temperatur.

Die spezielle Annahme über die Temperaturabhängigkeit der Aktivierungsvolumina ΔV und ΔW wurden dem Millerschen Ansatz entnommen. Als Versuchsergebnisse, die zur Bestimmung der Parameter herangezogen wurden, dienten Angaben von Miller [1], der für Reinstaluminium und die Stähle 304SS und 316SS eine größere Anzahl von Messungen stationärer Kriechgeschwindigkeiten in Abhängigkeit von Spannung und Temperatur zusammengestellt hat. Bild 4 zeigt ein Beispiel für den Stahl 304SS. Er benutzt diese Daten als Grundlage zur Aufstellung eines phänomenologischen Stoffgesetzes, das die Form

$$\dot{\varepsilon}_s = B\,\vartheta'(T)\left[\sinh(A\sigma)\right]^n \qquad \sigma, T \ldots \text{const.}$$

mit
$$\vartheta'(T) = \exp\left(-\frac{U^*}{0{,}6\,k_B T_M}\left(\ln(0{,}6\,T_M/T)+1\right)\right) \qquad \text{für } T < 0{,}6\ T_M \qquad (3.19)$$

$$\vartheta'(T) = \exp\left(-\frac{U^*}{k_B T}\right) \qquad\qquad \text{für } T \geq 0{,}6\ T_M$$

annimmt.

Für die Parameterbestimmung werden diese Beziehungen mit den von Miller ermittelten stoffabhängigen Konstanten als experimentelle Ergebnisse benutzt.Die Parameteranpassung erfolgt über die Beziehung

$$\int_G (\ln \dot{\varepsilon}_s - \ln \bar{\dot{\varepsilon}}_s)^2\, dG = \text{Min!} \qquad\qquad (3.20)$$

G ist dabei der Temperatur-Spannungsbereich über den die Anpassung vorgenommmen wird. Es werden logarithmierte Kriechgeschwindigkeiten verglichen, da dies gewährleistet, daß die Anpassung auch im Bereich niedriger Spannungen und Temperaturen, für die die Kriechgeschwindigkeiten sehr kleine Werte annehmen, gültige Vergleichsgrößen liefert.

Der Ausdruck

$$\ln \overline{\dot{\varepsilon}}_S = \underbrace{\ln(\lambda C_1) - \phi \ln(\frac{C_1}{C_2})}_{\tilde{A}} + \underbrace{\frac{1}{k_B}[(\phi-1)U^* - \phi \overline{Q}]}_{\tilde{B}} \frac{1}{T} +$$

$$+ \underbrace{\frac{\Delta \overline{V}}{k_B}(1-\phi)}_{\tilde{C}} \sigma \qquad (3.21)$$

mit

$$\phi = 1/(1 + \Delta \overline{W} / \Delta \overline{V})$$

für die anzupassende, stationäre Kriechgeschwindigkeit nach Gleichung (3.17) läßt sich in der Form

$$\ln \overline{\dot{\varepsilon}}_S = \tilde{A} + \tilde{B} \frac{1}{T} + \tilde{C} \sigma \qquad (3.22)$$

zusammenfassen. Man erhält also eine einfache Funktion der Variablen σ, $1/T$ und ein konstantes Glied.

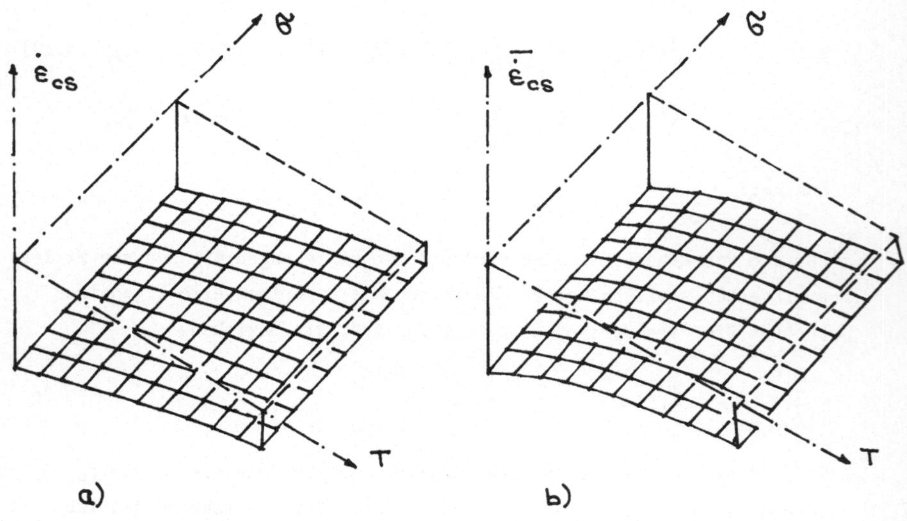

Bild 5: Stationäre Kriechgeschwindigkeiten für Reinstaluminium

    a) Ausgangsdaten nach Miller (Bild 4)

    b) Anpassung nach Gleichung (3.22)

Mit der Benutzung von Ergebnissen für stationäres Kriechen lassen sich
somit nur Aussagen über Kombinationen der Parameter in den Ansätzen
(3.1) bis (3.4) gewinnen. Dies erscheint verständlich, da sich bei
stationären Vorgängen die Verfestigungs- und Erholungsvorgänge teilweise
gegenseitig aufheben, und damit nicht erwartet werden darf, daß sich aus
derartigen Versuchen sämtliche Parameter des Modells eindeutig bestimmen
lassen.

Bild 5 zeigt die erzielte Anpassung der stationären Kriechgeschwindigkeiten an die Ergebnisse von Miller für 304SS. Die erhaltenen Werte für
die Parameterkombinationen $\tilde{A}$, $\tilde{B}$ und $\tilde{C}$ sind in Tabelle 1 angegeben.

	$\Delta\bar{W}/\Delta\bar{V}$	ϕ	$\Delta\bar{V}$	$\Delta\bar{W}$	Q^{*}
	–	–	mm^2/K	mm^2/K	N mm
Reinstaluminium	0,1	0,909	6,31 E-20	6,31 E-21	1,90 E-16
C1/C2 = 1	0,2	0,833	3,44 E-20	6,88 E-21	1,85 E-16
$300 \leq T \leq 700$ K	0,3	0,769	2,40 E-20	7,45 E-21	1,80 E-16
$10 \leq \sigma \leq 40$ N/mm^2	0,4	0,714	2,01 E-20	8,02 E-21	1,76 E-16
$\tilde{A}$ = 11,629	0,5	0,667	1,72 E-20	8,60 E-21	1,71 E-16
$\tilde{B}$ = -14123 K	0,6	0,625	1,53 E-20	9,17 E-21	1,66 E-16
$\tilde{C}$ = 0,4153 mm^2/N	0,7	0,588	1,39 E-20	9,74 E-21	1,61 E-16
kf = 112000	0,8	0,556	1,29 E-20	1,03 E-20	1,56 E-16
U^{*} = 2,43 E-16 Nmm	0,9	0,526	1,21 E-20	1,09 E-20	1,51 E-16
	1,0	0,500	1,15 E-20	1,15 E-20	1,47 E-16
304 SS	0,1	0,909	1,03 E-20	1,03 E-21	4,79 E-16
C1/C2 = 1	0,2	0,833	5,60 E-21	1,12 E-21	4,66 E-16
$500 \leq T \leq 1000$ K	0,3	0,769	4,04 E-21	1,21 E-21	4,52 E-16
$50 \leq \sigma \leq 150$ N/mm^2	0,4	0,714	3,27 E-21	1,31 E-21	4,38 E-16
$\tilde{A}$ = 13,080	0,5	0,667	2,80 E-21	1,40 E-21	4,24 E-16
$\tilde{B}$ = -35753 K	0,6	0,625	2,49 E-21	1,49 E-21	4,10 E-16
$\tilde{C}$ = 0,0676 mm^2/N	0,7	0,588	2,27 E-21	1,59 E-21	3,96 E-16
kf = 479000	0,8	0,556	2,10 E-21	1,68 E-21	3,28 E-16
U^{*} = 6,32 E-16 Nmm	0,9	0,526	1,97 E-21	1,77 E-21	3,68 E-16
	1,0	0,500	1,87 E-21	1,87 E-21	3,54 E-16

Tabelle 1: Parameter des Modells für Reinstaluminium und 304 SS

Tabelle 1 zeigt außerdem die Werte für die Modellparameter, die sich aus
diesen Ergebnissen bei Variation der Größe $\Delta\bar{W}/\Delta\bar{V}$ ergeben. Man erhält
Parameter, die in durchaus akzeptablen Größenordnungen liegen. Ihre
Zahlenwerte bestimmen das instationäre Verhalten des Modells.

Zur Untersuchung der Einflüsse der Parameterwahl auf instationäre Vorgänge wurden für eine Reihe von Parameterkombinationen, die jeweils auf die selben stationären Kriechgeschwindigkeiten führen, Transienten berechnet, die durch Spannungs- und Temperatursprünge entstehen.

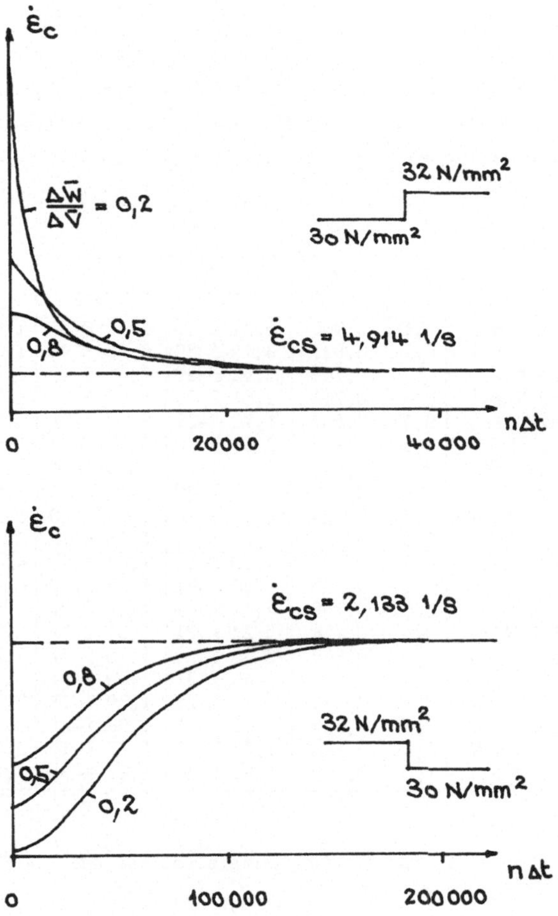

Bild 6: Transienten bei Reinstaluminium für Lasterhöhung und -absenkung

Dabei wurde jeweils für gegebene Ausgangsspannungen und -temperaturen die stationäre Verteilung und die zugehörige Formänderungsgeschwindig-

keit ermittelt, und danach ein Temperatur- oder Spannungssprung vorgegeben, der zu einer neuen stochastischen Matrix führt. Mit diesem verän-
derten System lassen sich mit Gleichung (3.11) die zeitliche Entwicklung
der Verteilung und die zugehörigen Formänderungsgeschwindigkeiten berechnen. Die Rechnung wurde soweit geführt, bis wieder die stationäre
Verteilung und stationäre Formänderungsgeschwindigkeit erreicht war.
Bild 6 zeigt Ergebnisse derartiger Berechnungen für Parameterkombinationen, die aus Anpassung an die experimentellen Ergebnisse von Miller für
Reinstaluminium ermittelt wurden.

Man erkennt, daß sich bei den Spannungs- und Temperatursprüngen typische
Transienten ausbilden, die sich in ihrem Verlauf grundsätzlich voneinander unterscheiden, je nachdem, ob es sich um einen Spannungs- oder
Temperaturanstieg oder um eine Verminderung dieser Werte handelt. Die
Übergänge auf den neuen stationären Zustand verlaufen bei einer Span-
nungs- oder Temperaturerhöhung nach einem starken Anstieg der Formänderungsgeschwindigkeiten erheblich rascher ab, als bei einer Verminderung der Spannung oder Temperatur. Im letzteren Fall ergibt sich eine
starke Erniedrigung und erst nach längerer Erholungszeit eine allmähli-
che Zunahme der Formänderungsgeschwindigkeit. Die Zeitspanne, die bis
zum Erreichen der neuen stationären Geschwindigkeiten verstreicht, ist
um ein Mehrfaches länger, als bei Spannungs- oder Temperaturerhöhung.

3.3 Erweiterung des Modells
3.3.1 Erweiterung des Zustandsraums
Die bisher dargestellten Ansätze zur Beschreibung der inelastischen
Formänderungsgeschwindigkeiten in Abhängigkeit von Spannung und Temperatur waren unter der Annahme aufgestellt, daß die Zahl der Fließeinheiten
oder Gleitmechanismen während der Vorgänge konstant bleibt. Für eine
Erweiterung des Modells, insbesondere auch zur Beschreibung des plastischen Verhaltens bei niedrigen Temperaturen ("spontane Plastizität") muß
diese Annahme aufgegeben werden. Um die mathematischen Hilfsmittel weiter nutzen zu können, die die Beschreibung der Vorgänge als Markov-
Prozesse zur Verfügung stellt, wird folgende Erweiterung des bisherigen
Modells angesetzt:

Das untersuchte System (Kontrollvolumen, materieller Körper) wird in

eine konstante Zahl von individuellen, voneinander unabhängigen Teilsys-
temen aufgeteilt gedacht, die jeweils homogenes Verhalten zeigen, d.h.
es wird angenommen, daß die inneren Spannungen, die den Verfestigungszustand kennzeichnen, für ein Teilsystem konstant sind. Ausserdem wird für
jedes Teilsystem eine bestimmte Dichte der Gleitmechanismen (= z.B.
einer Dichte von mobilen Versetzungen) vorausgesetzt, die sich ebenso
wie der Verfestigungszustand im Laufe des Vorgangs verändert.

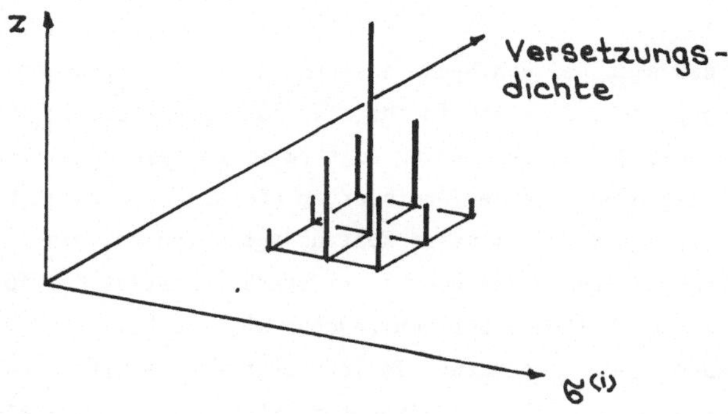

Bild 7: Erweiterter Zustandsraum

Diese Annahmen erfordern eine Erweiterung des Zustandsraumes auf eine
zweidimensionale Zustandsmannigfaltigkeit gemäß Bild 7, in der nun
Übergänge vom Ausgangszustand des Teilsystems zu den dargestellten Nach-
barzuständen möglich sind. Die Annahme einer konstanten Zahl von Teilsy-
stemen bedeutet jedoch, daß die zugehörige Matrix der Übergangswahrscheinlichkeiten weiterhin eine stochastische Matrix bleibt, und der
Vorgangsablauf durch eine Markov-Kette beschrieben werden kann. Da phy-
sikalische Argumente zur Verfügung stehen, die einen Teil der Übergänge
als wenig wahrscheinlich erscheinen lassen (z.B. besitzen die Zunahme
der Versetzungsdichte bei Erholung, oder die Abnahme der Versetzungsdichte bei Verfestigung, geringe Wahrscheinlichkeit), läßt sich die stochastische Matrix reduzieren. Als Extremfall kann an eine Beziehung
gedacht werden, die die Versetzungsdichte und die Größe der inneren
Spannungen koppelt, also zu einem funktionalen Zusammenhang der Art

$$S = S(\sigma, \dot{\omega})$$
(3.23)

führt. In diesem Falle läßt sich der Vorgang wieder im ursprünglichen
Zustandsraum beschreiben. Lediglich in den Beziehungen zur Ermittlung
der inelastischen Formänderungsgeschwindigkeiten ist ein Einfluß der
Versetzungsdichten zu berücksichtigen.

3.3.2 Entwicklungsgleichung für die Fließeinheitenanzahl

Beim Versuch der Anpassung der Konstanten an experimentelle Ergebnisse
von Servi und Grant [11] (diese liegen auch Millers Rechnungen zugrunde),Feltham [12,13] und Radhakrishnan [14] für Aluminium, Kupfer, Cobalt
und hochlegierten Chrom-Nickel-Stahl über Gleichung (3.22) zeigten sich
bei hohen Temperaturen teilweise erhebliche Differenzen zwischen den
gemessenen und den errechneten Logarithmen der Dehngeschwindigkeiten
(Bild 8).

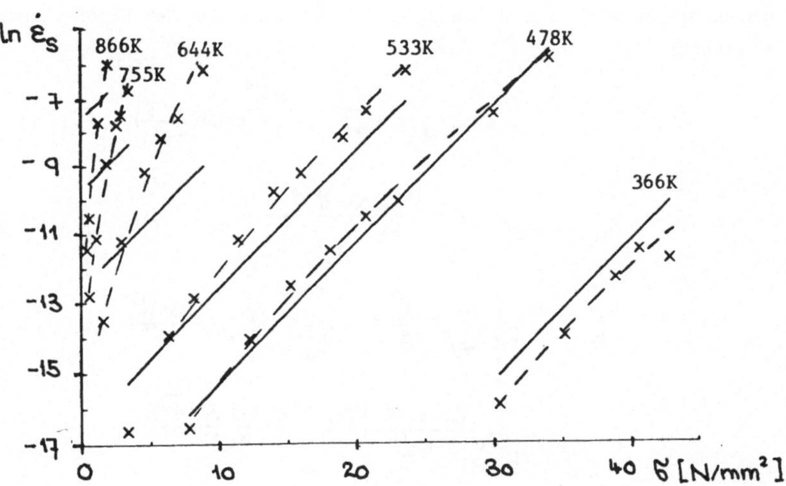

Bild 8: Stationäre Kriechdaten für Reinstaluminium nach [11]. Die
ausgezogenen Linien geben die Anpassung mit Gleichung (3.22) wieder.

Eine qualitative Auswertung der in den erwähnten Literaturstellen angegebenen Diagramme zeigt übereinstimmend, daß die Funktionen für $\ln \dot{\varepsilon}_s$ mit
wachsender Spannung einen leicht degressiven Anstieg aufweisen und insbesondere ihre Steigung bei gleicher Spannung mit wachsender Temperatur
zunimmt. Diese Eigenschaften sind in Gl.(3.22) nicht enthalten.

Von der Überlegung ausgehend, daß eine Verbesserung der Eigenschaften
des Modells zur Beschreibung realen Werkstoffverhaltens mit möglichst
einfachen Änderungen erreicht werden sollte, wurde für die praktische
Rechnung eine Erweiterung gemäß Gleichung (3.23) eingeführt.

Da es sich bei der Erzeugung von Versetzungen durch Frank-Read-Quellen
um einen mit der Hindernisüberwindung eng verbundenen Vorgang handelt,
lag es nahe, für die Erzeugung von Fließeinheiten gleichartige Beziehungen anzusetzen wie für die Beschreibung des Verformungsprozesses. Der
Gleichungssatz (3.1) bis (3.3) wurde dementsprechend wie folgt geändert
und um die Entwicklungsgleichung (3.27) für die Zahl der Fließeinheiten
erweitert:

$$\dot{V}_{i,i+1} = 2\,C_1\,\Delta t\,Z_{s,eff}\,e^{-\frac{U^*}{k_B T}}\,e^{-\frac{\Delta V \sigma^{(i)}}{k_B T}}\,\sinh\left(\frac{\Delta V \sigma}{k_B T}\right) \tag{3.24}$$

$$E_{i,i-1} = C_2\,\Delta t\,e^{-\frac{\alpha U^* - \beta\,\Delta V\,\sigma^{(i)}}{k_B T}} \tag{3.25}$$

$$\dot{\varepsilon} = 2\,C_1\,\lambda\,Z_{eff}\,e^{-\frac{U^*}{k_B T}}\,\sinh\left(\frac{\Delta V \sigma}{k_B T}\right)\sum_{i=1}^{\infty} z_i\,e^{-\frac{\Delta V \sigma^{(i)}}{k_B T}} \tag{3.26}$$

$$\dot{Z} = C_{P1}\,Z\left(e^{-\frac{U_q - \Delta V(\sigma - \bar{\sigma}^{(i)})}{k_B T}} - e^{-\frac{U_q + \Delta V(\sigma + \bar{\sigma}^{(i)})}{k_B T}}\right)$$

$$-C_{P2}\,Z\,e^{-\frac{\alpha' U_q - \beta'\,\Delta V\,\bar{\sigma}^{(i)}}{k_B T}} \tag{3.27}$$

$$= Z\left(C_{P1}\,e^{-\frac{U_q + \Delta V\,\bar{\sigma}^{(i)}}{k_B T}}\,2\,\sinh\left(\frac{\Delta V \sigma}{k_B T}\right) - C_{P2}\,e^{-\frac{\alpha' U_q}{k_B T}}\,e^{-\frac{\beta'\,\Delta V\,\bar{\sigma}^{(i)}}{k_B T}}\right)$$

Hierbei sind

C_{P1}, C_{P2} ... Raten für Fließeinheitenproduktion bzw. -auflösung

U_q ... Aktivierungsenergie für Fließeinheitenproduktion

$\alpha' U_q$ ... Aktivierungsenergie für Fließeineiheitauflösung

$\beta' \Delta v$ ... Aktivierungsvolumen für Fließeinheitenauflösung

Die für die inelastischen Formänderungsgeschwindigkeiten wirksame Zahl der Fließeinheiten ergibt sich aus den Beziehungen

$$Z_{eff} = Z - Z_{0,S} \left(\frac{T}{T_M}\right) \; ; \; Z_{s,eff} = Z_s - Z_{0,S} \left(\frac{T}{T_M}\right) \qquad (3.28)$$

hierbei gilt Z_{eff} für Transienten, $Z_{s,eff}$ für stationäre Vorgänge. $Z_{0,S}$ ist dabei die minimale Fließeinheitendichte, die auch bei beliebig langer Erholung des Werkstoffs bei der Temperatur T nicht unterschritten wird. $\overline{\aleph^{(i)}}$ stellt die innere Spannung dar, die dem Mittelwert der Verteilung zugeordnet ist.

Der Verformungsprozess soll nur von der effektiven Fließeinheitenzahl abhängen. Die immer im Werkstoff enthaltene Grundversetzungsdichte soll keinen direkten Einfluß auf die Verfestigung besitzen. Das Aktivierungsvolumen für die Fließeinheitenproduktion soll dem der Hindernisüberwindung gleich sein, da es sich um einen damit eng gekoppelten Vorgang handelt. Alle übrigen Größen in der Beziehung für die Änderung von Z wurden zunächst gegenüber denen des Verformungsprozesses als verschieden angesetzt. Es ist jedoch wahrscheinlich, daß sich im weiteren Verlauf der Arbeiten zeigen wird, daß auch andere Größen in beiden Vorgängen gleichartig angesetzt werden dürfen.

Um das mathematische Modell zunächst möglichst einfach zu halten, wurde angenommen, daß die Fließeinheitenerzeugung nicht von der Verteilung der Fließeinheiten über den inneren Spannungen, sondern nur von der Lage des Mittelwerts dieser Verteilung abhängen.

Berechnet man für den erweiterten Gleichungssatz wiederum die stationäre Dehngeschwindigkeit, so ergibt sich

$$\ln \dot{\varepsilon}_s = \ln(2C_1\lambda) - \frac{1-\beta}{1+\beta}\ln(\frac{2C_1}{C_2}) + \frac{\beta}{1+\beta'}\ln(\frac{2C_{P1}}{C_{P2}}) \qquad (3.29)$$

$$- \frac{1}{k_BT}(\alpha\, U^* - \frac{\Delta V \Delta V^{(i)}\beta}{2(1+\beta)} + \frac{U^*\beta(1-\alpha')}{1+\beta'}) + \frac{\beta}{1+\beta'}\ln(\sinh(\frac{\Delta V \vartheta}{k_BT})).$$

Setzt man das Aktivierungsvolumen als linear von der Temperatur abhängige Größe an, so wird daraus

$$\ln \dot{\varepsilon}_s = k_1 - \frac{U^*}{k_BT}k_2 + k_3 \ln(\sinh(\frac{k_4\,\vartheta}{k_B})) \qquad (3.30)$$

Dies entspricht der aus vielen empirischen Arbeiten bekannten Beziehung für das stationäre Kriechen bei der $\dot{\varepsilon}_s$ von der Potenz des $\sinh(\vartheta)$ abhängt. Hier konnte diese Gleichung zwanglos aus den anfangs gemachten Modellvorstellungen abgeleitet werden.

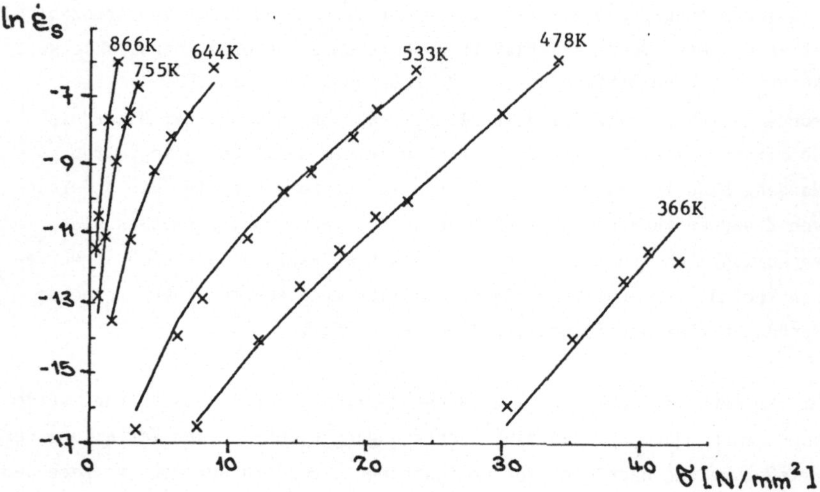

Bild 9: Stationäre Kriechgeschwindigkeiten nach]11]. Die ausgezogenen Linien stellen die Anpassung nach Gleichung (3.31) dar.

Werkstoff	U^a [kJ/mol]	Anzahl der Meßwerte	Wertebereich		angepaßte Werte für				mittlere Abweichung in % zwischen $\ln \dot{\varepsilon}_R$ und $\ln \dot{\varepsilon}_M$	Quelle
			σ [N/mm^2]	T [K]	k1	k2	k3	k4		
Al 99.99	149	38	0.44– 42.9	366–866	25.17	1.09	3.8	0.371	2.11	[11]
Cu	200	67	11.2 –100.7	673–823	14.54	0.77	3.77	0.103	3.3	[12]
Cu/Zn 90/10	200	43	13.0 – 97.0	723–923	21.14	0.97	4.07	0.067	1.6	[16]
Cu/Zn 80/20	200	24	13.0 – 97.0	673–873	23.34	0.86	4.19	0.027	1.6	[16]
Cu/Zn 70/30	200	35	12.0 –105.0	673–823	19.29	0.86	3.53	0.085	3.1	[16]
Cu/Zn 65/35	200	26	11.0 – 98.0	623–823	36.25	0.81	3.90	0.001	2.0	[16]
X1CrNi18 9	280	13	235.0 –392.0	773–873	20.10	0.91	6.30	0.028	1.6	[14]
X8CrNi16 13	280	22	63.0 –305.0	873–1023	44.10	2.04	1.34	0.382	3.7	[17]
Co	275	37	25.0 –205.0	773–1023	40.88	1.00	5.66	0.005	2.7	[13]
W	640	40	1.6 –141.0	1873–3073	22.89	0.95	4.01	0.440	4.3	[15]
Pb Korn Ø 0.04	110	10	6.5 – 12.5	298–372	25.78	1.09	2.83	1.064	2.8	[18]
Pb Korn Ø 0.06	110	8	5.0 – 12.5	298–368	27.37	1.07	5.20	0.600	1.3	[18]
Pb Korn Ø 0.11	110	9	6.5 – 10.0	298–353	28.64	1.10	3.00	1.000	3.7	[18]
Al Korn Ø 0.25	149	15	24.0 –137.0	253–531	39.21	0.85	1.00	0.034	4.9	[19]
Al/Mg 95/ 5	149	10	6.7 – 18.9	623–673	26.18	1.11	3.50	0.091	0.4	[20]

Tabelle 2: Werte der freien Parameter $k_1 \ldots k_4$ in Gl. (3.31) für verschiedene Werkstoffe

Für eine Reihe von Meßdaten für verschiedene Metalle und Legierungen
wurden Anpassungen der Parameter k_1 bis k_4 vorgenommen. Die Ergebnisse
waren durchaus zufriedenstellend. Es zeigte sich jedoch, daß hinsichtlich der Genauigkeit der Wiedergabe der gemessenen Werte gegenüber
Berechnungen mit Gl. (3.29) eine weitere Verbesserung erzielt werden
kann, wenn ΔV als temperaturunabhängige Größe angesetzt wird. Das Argu-
ment des sinh() ist dann eine Funktion von σ/T :

$$\ln \dot{\varepsilon}_\sigma = k_1 - \frac{U^*}{k_B T} k_2 + k_3 \ln(\sinh(\frac{k_4 \sigma}{k_B T})). \tag{3.31}$$

In Bild 9 erkennt man die Verbesserung gegenüber früheren Ergebnissen
(Bild 8). Die erzielten Genauigkeiten bei der Erfassung von Meßwerten
für unterschiedliche Werkstoffe sind in Tabelle 2 zusammengefaßt. Übereinstimmend zeigt sich, daß mit dieser Art von Gleichung eine gute
Wiedergabe der Meßwerte erreicht wird. Bemerkenswert ist dabei, daß die
Werte der sich entsprechenden Parameter bei allen Werkstoffen etwa in
der gleichen Größenordnung liegen. Dies spricht dafür, daß die zugrunde
liegenden Ansätze auch physikalisch sinnvoll sind.

4 Zusammenfassung

Die Behandlung von Kriechen und Plastizität mit Hilfe des mathematischen
Modells einer Markov-Kette stellt Berechnungsmethoden zur Verfügung, bei
denen sich durch die Übergangswahrscheinlichkeiten in der stochastischen
Matrix und die Verteilung der "Fließeinheiten" über der mit inneren
Spannungen identifizierten Zustandsachse die Struktur des Systems und
deren Veränderung in Abhängigkeit von Temperatur und äußerer Spannung in
übersichtlicher Weise beschreiben läßt. Das Modell besitzt inhärent
typische Eigenschaften, die für Kriechvorgänge kennzeichnend sind (stationäre Zustände bei homogenen Prozessen, typische Verläufe der Transienten bei Last- und Temperaturwechseln). Die Berechnung der stationären
Zustände, der zugehörigen Formänderungsgeschwindigkeiten und der Verän-
derungen des Systems bei Lastwechseln erfolgt dabei streng, d.h. die in
das Modell eingebrachten Informationen bleiben auch bei beliebig weit
geführter Vorgangsbeschreibung erhalten.

Das Modell wurde mit Hilfe von in der Theorie der Hochtemperaturplastizität wohlbekannten Ansätzen über die Grundmechanismen, die zu inelasti-

schen Formänderungen führen (thermisch aktivierte Platzwechsel mit Unterstützung durch äußere Spannungen), formuliert. Mit diesen Mechanismen lassen sich Strukturänderungen verknüpfen, die Verfestigung des Werkstoffs zur Folge haben. Zusätzlich wurden Erholungsmechanismen eingeführt, die ebenfalls als thermisch aktivierte Vorgänge angesetzt wurden, die von den mit der Verfestigung verbundenen inneren Spannungen unterstützt werden.

Untersuchungen zur numerischen Bestimmung der Modellparameter, die wegen der Verfügbarkeit von experimentellen Daten zunächst für stationäre Vorgänge durchgeführt wurden, zeigen, daß mit wenigen Parametern eine Beschreibung experimenteller Ergebnisse über weite Temperatur- und Spannungsbereiche möglich ist, und daß die sich bei diesen Anpassungen ergebenden Beziehungen bekannten empirischen Kriechansätzen entsprechen, wobei sie jetzt aber auf physikalisch begründeten Ansätzen aufbauen. Wie zu erwarten ist, lassen sich bei der Heranziehung stationärer Kriechversuche jedoch nur Größen quantitativ bestimmen, die mehrere Ansatzparameter zusammenfassen. Der vollständigen Parameterbestimmung müssen instationäre Vorgänge zugrundegelegt werden, da sich im stationären Fall die einzelnen Mechanismen teilweise kompensieren und sie damit aus dem Experiment nicht eindeutig identifiziert werden können.

Schrifttum

[1] Miller, A.K.: A Unified Phenomenological Model for the Monotonic, Cyclic and Creep Deformation of Strongly Workhardening Materials. Diss., Stanford University May 1975

[2] Valanis, K.C., Lee, C.F.: Deformation Kinetics of Steady-State Creep in Metals. Int. J. Solids and Structures, Vol. 17, pp. 589-604, 1981

[3] Lubliner, J.: On the Thermodynamic Foundations of Non-Linear Solid Mechanics, Int. J. Non-Linear Mechanics, Vol.7, pp. 237-254, 1972

[4] Tokuoka, T.: Thermodynamical Deduction of Constitutive Equations of a Plastic Material with Combined Work-Hardening. Acta Mechanica,

Vol. 37, pp. 191-198, 1980

[5] Mc Cartney, L.N.: Constitutive Relations Describing Creep Deformation for Multi-Axial Time-Dependent Stress State. J. Mech. Phys. Solids, Vol. 29, pp. 13-33, 1980

[6] Täubert, P.: Eine Kriechtheorie für Metalle unter Berücksichtigung von Verfestigung und Erholung. Abhandlungen der Deutschen Akademie der Wissenschaften zu Berlin, Klasse Mathematik, Physik und Technik. Jahrgang 1958, Nr. 7, Berlin

[7] Feller, W.: An Introduction to Probability Theory and Its Applications. Vol.1, 2d. ed., John Wiley & Sons, New York, 1957.

[8] Karlin, S. and H.M. Taylor: A Second Course in Stochastic Processes. Academic Press, New York, 1981.

[9] Bharucha-Reid, A.T.: Elements of the Theory of Markov-Processes and Their Applications. Mc Graw-Hill Book Co. New York, 1960.

[10] Dorn, J.E.: Creep and Recovery, p.274, American Society for Metals Metals Park, Ohio 1957

[11] Servi, I.S. and N.J.Grant: Creep and Stress Rupture Behavior of Aluminum as a Function of Purity. J. of Metals. Trans. AIME 191 (1951) 909-916

[12] Feltham, P. and Meakin, J.D.: Creep in Face-Centered Cubic Metals with Special Reference to Copper. Acta Metallurgica, Vol. 7, (1959),pp. 614-627

[13] Feltham, P. and Myres, T.: Energy Barriers Controlling Creep Prozesses in Metals with Special Reference to ß-Cobalt. Philosophcal Magazine, 8, (1963), pp. 203-211

[14] Radhakrishnan, V.M.: Untersuchungen zum Verhalten warmfester Stähle unter veränderlicher mechanischer Beanspruchung im Zeitstandbereich.

Arch. Eisenhüttenwesen 46 (1975) Nr.5, S. 335-339

[15] Sherby, O.D. and Miller, A.K.: Combining Phenomenology and Physics
in Describing the High Temperature Mechanical Behavior of Crystal-
line Solids. J. of Engineering Materials and Technology, 101,
(1979), pp. 387-395

[16] Feltham, P. and Copley, G.J.: Creep in Face-centred Metals and
Solid Solutions with Special Reference to -Brasses. Phil. Mag.
Ser 8, 5, (1960), S. 649ff

[17] Schmidt, W. und v.d.Steinen, A.: Beitrag zum Einfluß einer Unter-
brechung der mechanischen Beanspruchung auf das Kriechverhalten
eines hochwarmfesten Stahles. Arch. Eisenhüttenwesen, 44, (1973),
Nr.4, S. 291-296

[18] Feltham, P.: On the Mechanism of High-Temperature Creep in Metals
with Special Reference to Polycristalline Lead. Proc.Phys.Soc. B69,
(1956), pp. 1173-1188

[19] Luthy, H., Miller, A.K.; Sherby, O.D.: The Stress and Temperature
Dependence of Steady-State Flow at Intermediate Temperatures for
Polycristalline Aluminum. Acta Metallurgica 28 (1980) 169-178

[20] Ahlquist, C.N. and W.D. Nix: The Measurement of Internal Stresses
During Creep of Al-Mg Alloys. Acta Metallurgica 19 (1971) 383.
Tab.3

The Industrial Use of Computer Programs in Cold Forging

JF. RENAUDIN - CETIM Saint-Etienne / FRANCE
HJ. BRAUDEL - CEMEF Sophia-Antipolis / FRANCE

Summary

Double simultaneous extrusion has been used in cold forging as a way to
simplify the forming of components with complex shapes. Simulation programs
for predicting simultaneous extrusions are ready to use as a part of CAD fcr
the cold extrusion of metals.

The programs are suitable for axisymmetric components produced by :
- tubular extrusion
- can with can extrusion
- forward (rod) - backward (can) extrusion

The present paper presents the work carried out by both CETIM and CEMEF. The
method that was chosen is based upon the upper bound method with optimization. The programs have been used to compare theory and practice. Good
agreement is claimed.

In future the programs may be extended to account for non-axisymmetric
shapes or other forming processes as rolling.

0   Introduction

For a few years, theoretical studies have been engaged in CEMEF* a laboratory of a French Technical University in order to give a precise enough
theoretical method to predict and simulate numerically the double cold
extrusion of steel.

This work has been sponsorshiped by CETIM, who evaluated experimentally
the validity of the method and made the numerical simulation ready to be
used as a part of a Computer Aided Design in the cold extrusion of steel.

(*) CEMEF : Centre de mise en forme des matériaux - Ecole des Mines de PARIS

The result of this cooperation has been a set of three programs which simlulate :
- simultaneous rod and can extrusion
- can with can extrusion
- tubular extrusion

For some monthes this research cooperation has been stopped and we are at the present time when these CAD programs are transfered towards the industry.

1   Available Geometries

As it will be hereafter recalled, the method which was chosen to predict simultaneously the extrusion-forces and the end-geometry of the components is based on the upper-bound method.

A consequence of this choice is theoretically that every new geometry of the tools should need a specific velocity-field.

At the presend time, only axisymetric geometries are available for the following operations :
- Backward - forward extrusion
- Can - with - can extrusion
- Tubular extrusion

Figures 1 give the geometrical input-parameters that must be introduced by the designer.

Some of these parameters are determined by the end design as soon as the remaining ones can be freely chosen by the designer in order to optimize the process. As an example, the designer can let the width of the punch-land (E_{poin}) vary in order to choose the optimal value E_{poin}^{*} that will give a more exact can length.

2   Method

The theory of the present method has been developped by CEMEF. It is based on the upper bound method and has been detailed in some publications (1) to (4) and two thesis (5) (6).

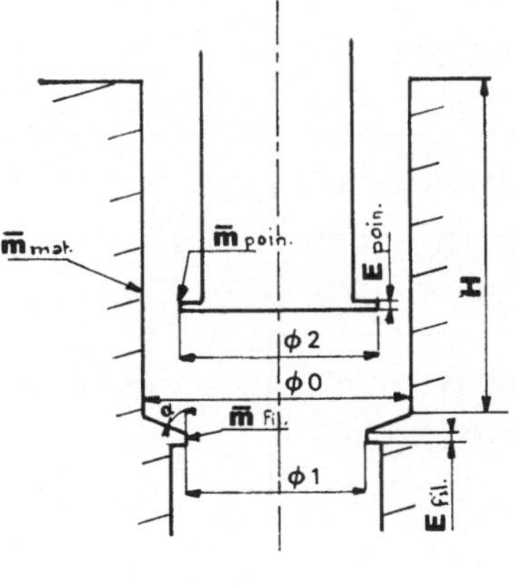

Fig 1a

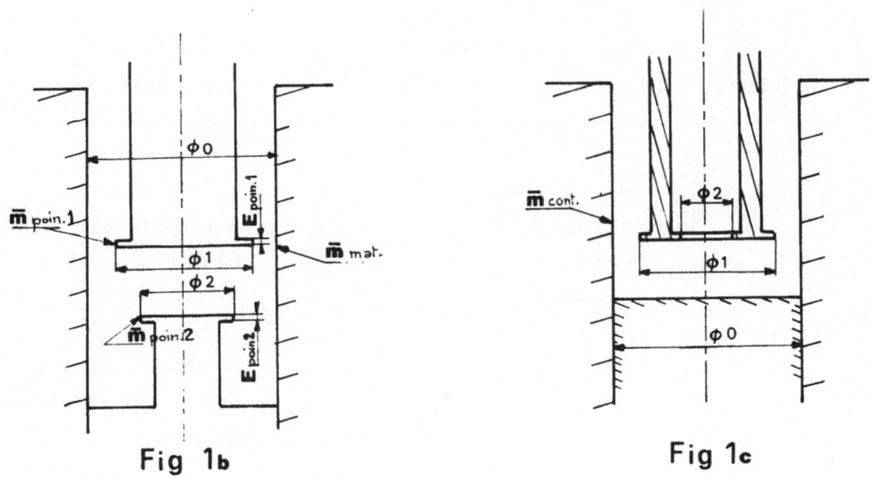

Fig 1b   Fig 1c

Figure 1 : Axisymetric simultaneous extrusions
  1a/ Backward and forward extrusion
  1b/ Can with can extrusion
  1c/ Tubular extrusion

2.1 Development_of_the_method (from (4))

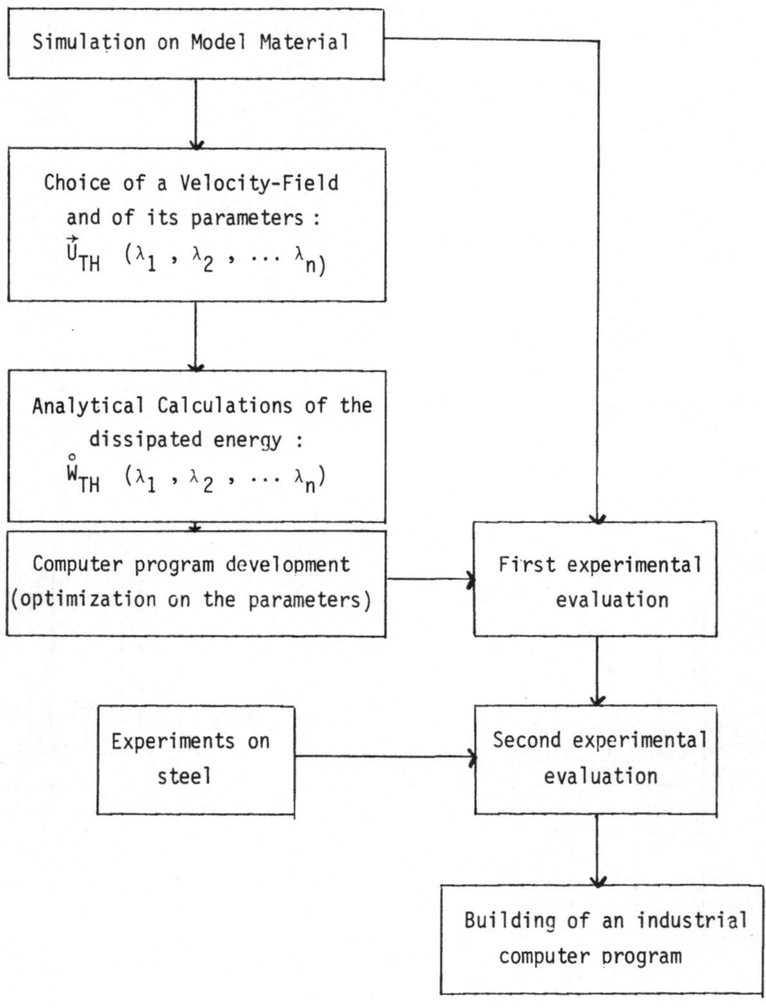

2.2 Description of the method

The stroke of the press in devided into a freely chosen number of steps. At each step of the stroke, the calculations start from the evaluation of the geometry and the material state that has been done at the former step.

A velocity field has been chosen which depends on n parameters, $\lambda_1,\ldots\lambda_n$. At each step the computer determines the parameters $\lambda_1^*,\ldots\lambda_n^*$ which minimize the total deformation power (the total deformation power is due to internal shear, deformation and friction). From the upper bound theory it is however known that the corresponding power is greater than the dissipated power in the real forming.

By identifying these powers an evaluation of the real but unknown extrusion pressure can be done.

The evaluation of the geometry is done when indentifying the real but unkown velocity field to the analytical one which is defined by the optimal parameters $\lambda_1^*,\ldots \lambda_n^*$.

The use of the upper bound method assumes that the material is not hardenable. The transformation of the mechanical properties of the material has thus been taken into occount after each step of the stroke in each zone of the velocity field (average strain-hardening).

2.3 Choice of the velocity-field

The simulation of the extrusion by model material has brought the elements allowing the building of a realistic velocity-field. For instance on figure 2 the beginning of a backward-forward extrusion has been simulated. The different zones that can be seen in the reality are represented. A velocity-field has been built, taking into account the boundaries between these zones. The optimization parameters have been chosen in this case as the angles α_{1s} and α_{1i} and the value λ. λ is defined as the ratio between the velocity of the rigid zone and that of the punch.

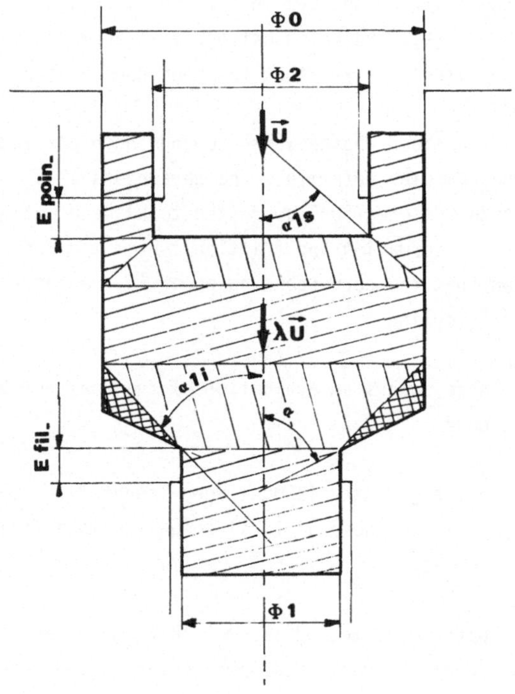

Figure 2 : Calculation-zones at the beginning of a simultaneous
rod and can extrusion

3   Mechanical properties of the metal

In order to describe the work-hardening of the metal, the strain-stress
law is assumed to have the following form (Krupkowski's law) :

$$\sigma_0 = \sigma_{00} \, (\bar{\varepsilon}_0 + \bar{\varepsilon})^n$$

where $\sigma_0 = \sigma_0 \, (\bar{\varepsilon})$ is the yield stress

The values σ_{00}, $\bar{\varepsilon}_0$ and n can be obtained very easily from a simple mechanical test such as a tension test.

It is therefore not a problem for the extruder to have these data and to
build a library of them for his most usual materials.

4   Sliding between workpiece and tools

It has been seen that the upper bound method was used in the program. Thus
its authors were obliged to use the friction factor $\bar{m}$ to simulate the
sliding (5) (6) :

$$\tau_i = \bar{m} \, \frac{\sigma_0}{\sqrt{3}} \quad , \quad \text{where}$$

τ_i is the shear stress at the boundary between workpiece and tool.
   (τ_i is assumed to be constant)
σ_0 is the yield stress of the forged material.

Usually the values of $\bar{m}$ are obtained from the ring-test or by upsetting
a cylinder.

These both tests show however a type of deformation which is not always
representative of an extrusion. Indeed the local surface deformation is
sometimes far more greater in an extrusion. For instance it is a nonsense
to assume $\bar{m} = .15$ in a can with can extrusion at the extenal point situated
at the web level (point A in fig. 3) where the surface extension can be
greater than 15 !

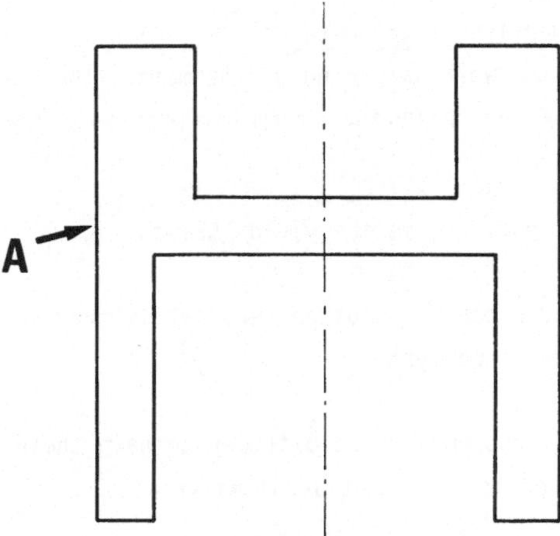

Fig. 3 : Can with can extrusion

Taking into account this difficulty, the program needs from the operator only three average friction-coefficients that are assumed constant along the forming operation :

- $\bar{m}_{poin}$ between the workpiece and the punch
- $\bar{m}_{fil}$ between the workpiece and the die-insert (or counterpunch)
- $\bar{m}_{cont}$ between the workpiece and the container

5   Description of the CAD-Program

5.1   Input of the datas

The input-datas are given to the computer in an interactive way. Figure 4a gives an image of the terminal screen during the input. These input-datas are the following ones :
- The values of the geometrical parameters, as shown on figures 1
- The values of the Tresca friction coefficient $\bar{m}$ for :
  . the sliding between punch and metal
  . the sliding between container and metal
  . the sliding between die and metal
- The mecanical properties of the metal : σ_{oo}, $\bar{\varepsilon}_o$, n
  that are used in the Krupkowski's law $\sigma_o = \sigma_{oo} (\bar{\varepsilon}_o + \bar{\varepsilon})^n$

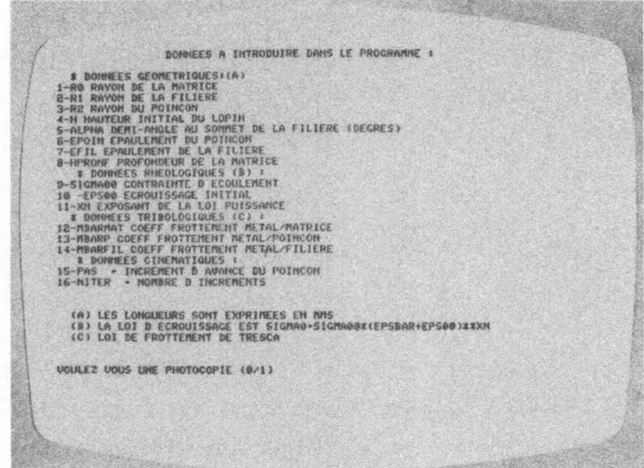

Fig 4a

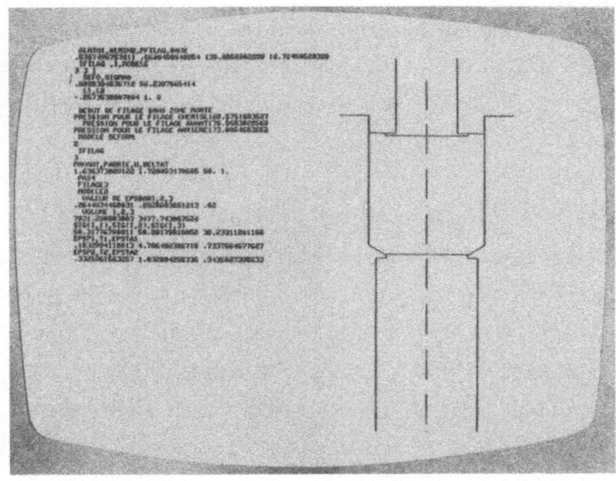

Fig 4b

Figure 4 : Simultaneous rod and can extrusion (FILMIX)

    4a/ Input of the datas

    4b/ Display of the internal parameters

5.2 Simulation of the extrusion

With the mechanical model that has been chosen, it has been theoretically
demonstrated (6) that only elementary processes occur during the early
stage of backward and forward extrusions. For example, let us consider a
"simultaneous" backward (can) and forward (rod) extrusion. The instantaneous
material flow is proved to be either a pure rod extrusion, a pure can
extrusion, or a rod extrusion without any sliding between the workpiece and
the container.

This property is in accordance with the reality, that has been observed
experimentally on model material and on steel.

It is possible to follow the modelised extrusion on the screen at each step
of the stroke, as shown on figure 4b .One can there fore have a real simulation of the material flow during the extrusion. Figures 5 give an example
of this simulation at the very beginning of the stroke.

The extrusion begins as a pure forward extrusion (fig. 5a to 5c). It is
followed by a pure backward extrusion (fig. 5d) and then by a rod extrusion
until the end of the stroke (fig. 5e to 5g) without any relative displacement
at the container interface.

However the choice that is made by the computer between the different
kinds of flow is done on the numerical values of the deformation power
that are obtained for each kind of extrusion. As a consequence a very and
physically unsignificant difference on an input data can induce in some
very scarce cases different kind of flow. If such an instability is observed
the knowledge of the internal parameters can be obtained at each step of the
stroke (fig. 4b).

In most cases, the only end-result (fig. 6) is needed.The simulation is
translated into graphical curves (fig. 7), which represent the evaluation
of the relation between extrusion lengths and stroke (fig. 7a) or extrusion
pressure and stroke (fig. 7b).

A possibility of comparison with other for instance experimental results
can been done on the screen.

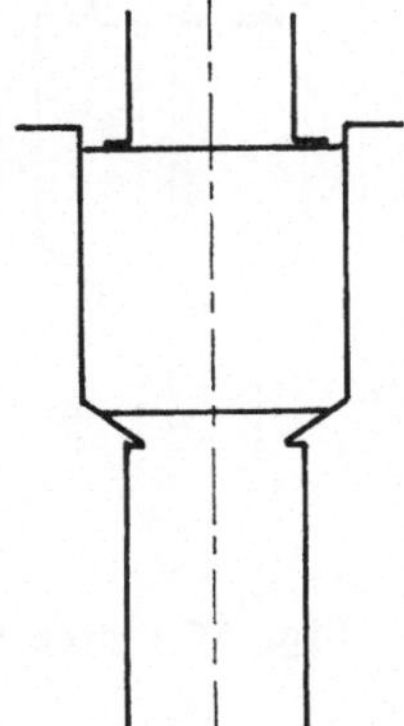

Fig. 5a : step 1

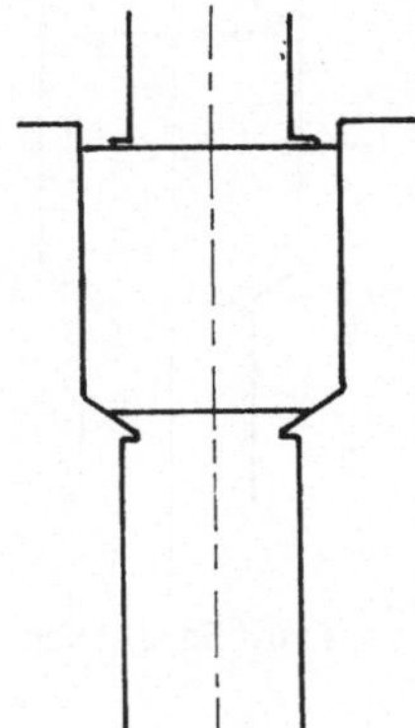

Fig. 5b : step 2

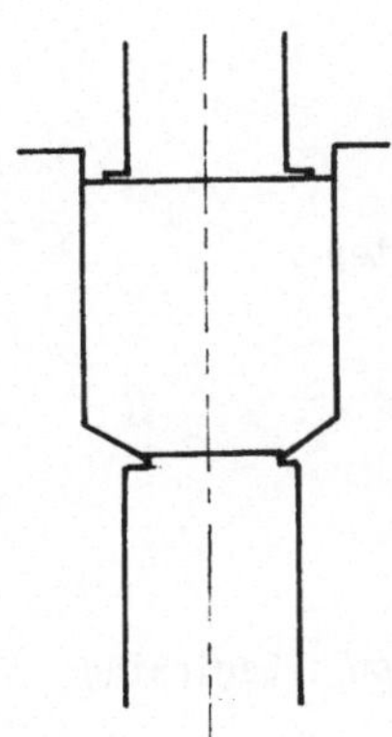

Fig. 5c : step 1

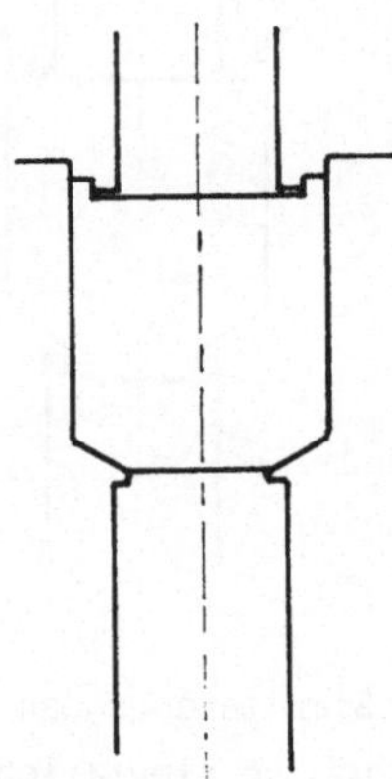

Fig. 5d : step 4

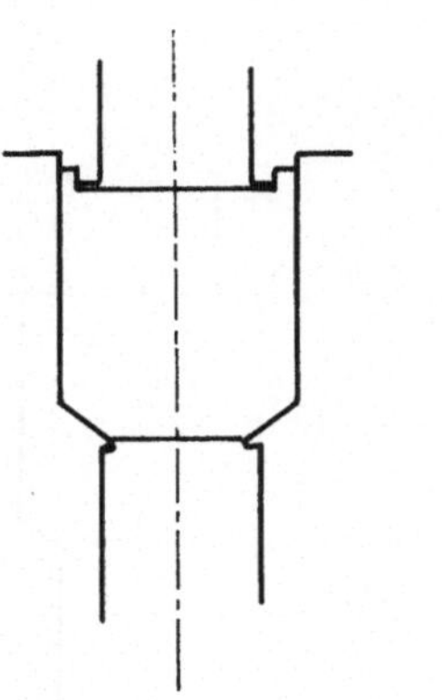

Fig. 5e : step 5

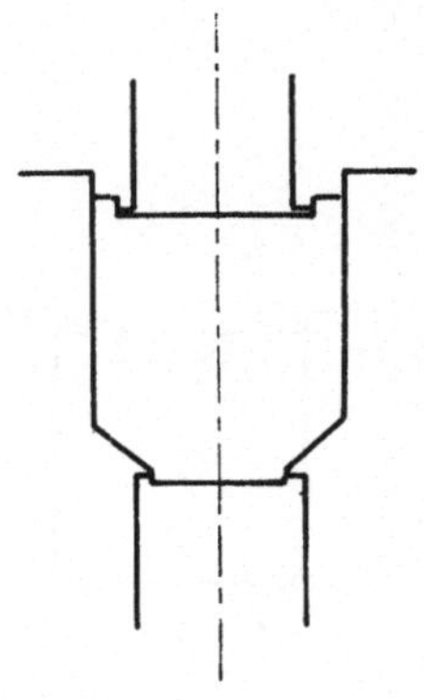

Fig. 5f : step 6

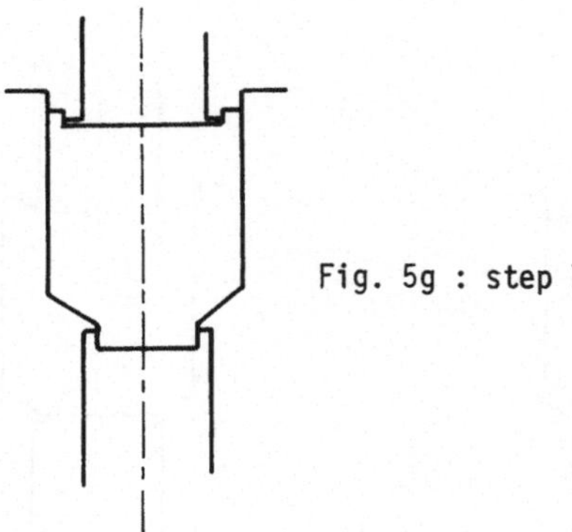

Fig. 5g : step 7

Figure 5 : Simultaneous can and rod extrusion ; Beginning
of the simulation

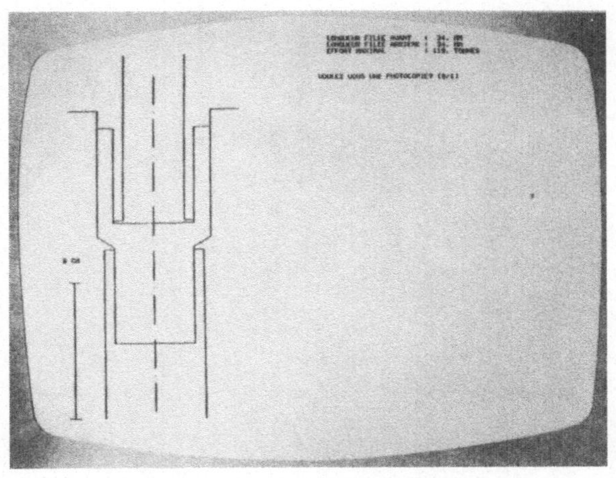

Figure 6 : Simultaneous rod and can extrusion (FILMIX)
Display of the end-results

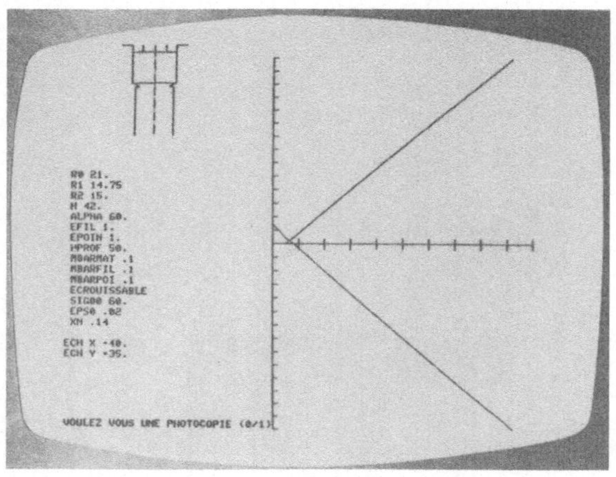

Fig 7a

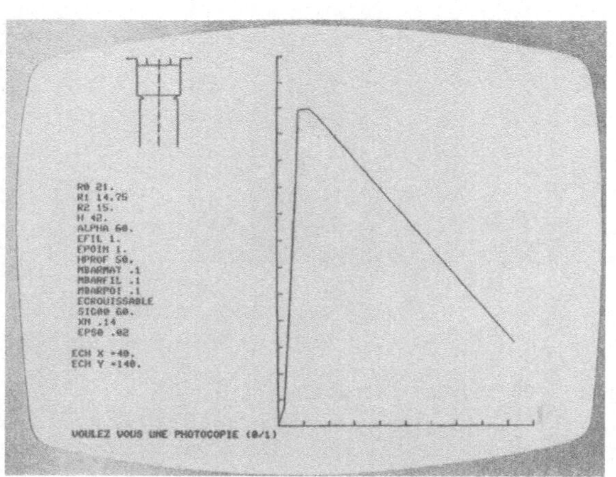

Fig 7b

Figure 7 : Simultaneous rod and can extrusion (FILMIX)
7a/ Extruded lenghts versus stroke
7b/ Extrusion-force versus stroke

5.3 Parametrical studies

It has been shown that the program can give a simulation of the whole
extrusion. Most of the time when a parametrical study is done, only the end
results are needed. Figure 8 represents these results as obtained on a
printer.

At the present time the parametrical study can be done on single parameters
and the optimization must be done by the designer. However this version has
been built in such a way that it could be very easily inserted into a
whole CAD program that would contain a multi-parameter optimization modulus.

6 Experimental evaluation of the methods

The principles of the method has been experimentally evaluated by comparing
the results of the program with real extrusions on steel in case of simultaneous rod and can extrusion. As the same method has been used for can
with can and tubular extrusion, these programs have been evaluated by tests
on model material or comparison with results of the litterature (8).

In order to verify the forecast that was given by the program, experimental
extrusions have been done. 30 different values of the parameters were chosen
on a backward-forward simultaneous cold extrusion. Table 1 gives the experimental plan.

The experiments were donc on a hydraulical press of 3000 KN. The ram velocity
was 15 mms^{-1}.

The material of the component was a XC 10 Steel (0,10 % of carbon), the
stress-strain curve of which is given in figure 9.

6.1 Component geometry

The comparison between experimental results and calculated datas is shown
on figure 10 on an example. The correlation is pretty good.

```
FILAGE                              ****************
                                    *              *
                                    *  CAS NO :    *
                                    *              *
PROGRAMME CETIM DE SIMULATION       ****************
DU FILAGE AVANT-ARRIERE             *              *
                                    *   PAGE  1    *
                                    *              *
                                    ****************
```

RO (MM)	R1 (MM)	R2 (MM)	ELANCE (MM)	EPOIN (MM)	EFIL (MM)	HPRONF (MM)
21.00	13.25	15.00	42.00	1.00	1.00	50.00

ALPHA (DEG)	SIGMAOO (MPA)	EPSOO	N	MBARMAT	MBARP	MBARFIL
60.00	600.0	.02	.14	.10	.10	.10

```
EFFORT MAXIMAL DE FILAGE      :   1.54 MN

LONGUEUR FILEE AVANT          :   41.0 MM

LONGUEUR FILEE ARRIERE        :   34.0 MM
```

Figure 8 : Simultaneous can and rod extrusion ; Printed results

Filières $\left\{\begin{array}{l} \text{Ø (mm)} \\ 2\alpha \\ \%\ R_A \end{array}\right.$	Ø 29,5 / 90° / 50 %	Ø 29,5 / 120° / 50 %	Ø 26,5 / 60° / 60 %	Ø 26,5 / 90° / 60 %	Ø 26,5 / 120° / 60 %	Ø 26,5 / 160° / 60 %	Ø 23 / 90° / 70 %	Ø 23 / 120° / 70 %	Ø 19 / 90° / 80 %	Ø 19 / 120° / 80 %
Poinçons Ø = 26,5 mm ; % R_R = 40 %			▨	▨	▨	▨	▨	▨	▨	▨
Ø = 30 mm ; % R_R = 50 %	▨	▨	▨	▨	▨	▨	▨	▨	▨	▨
Ø = 35 mm ; % R_R = 70 %	▨	▨	▨	▨	▨	▨	▨	▨	▨	▨
Ø = 37 mm ; % R_R = 78 %									▨	▨

Table 1 : Simultaneous can and rod extrusion ; experimental plan

     a   : Die-insert

     b   : Punch

   ▨   : Experimental shapes

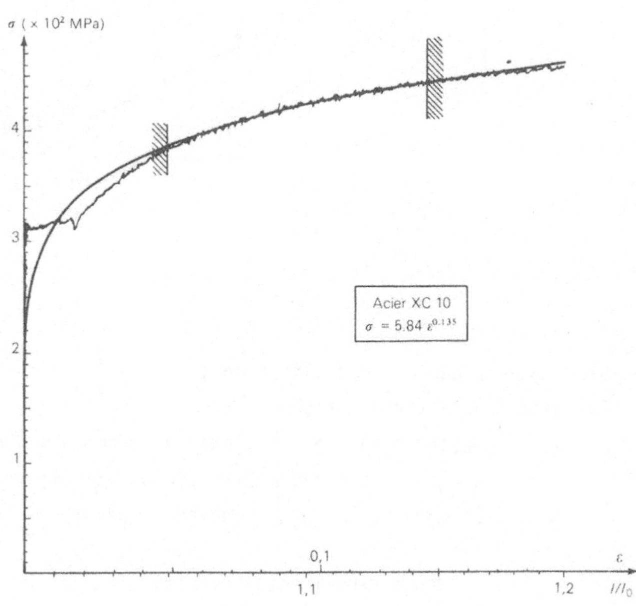

Figure 9 : XC 10 Steel ; Stress-strain curve (tension test)

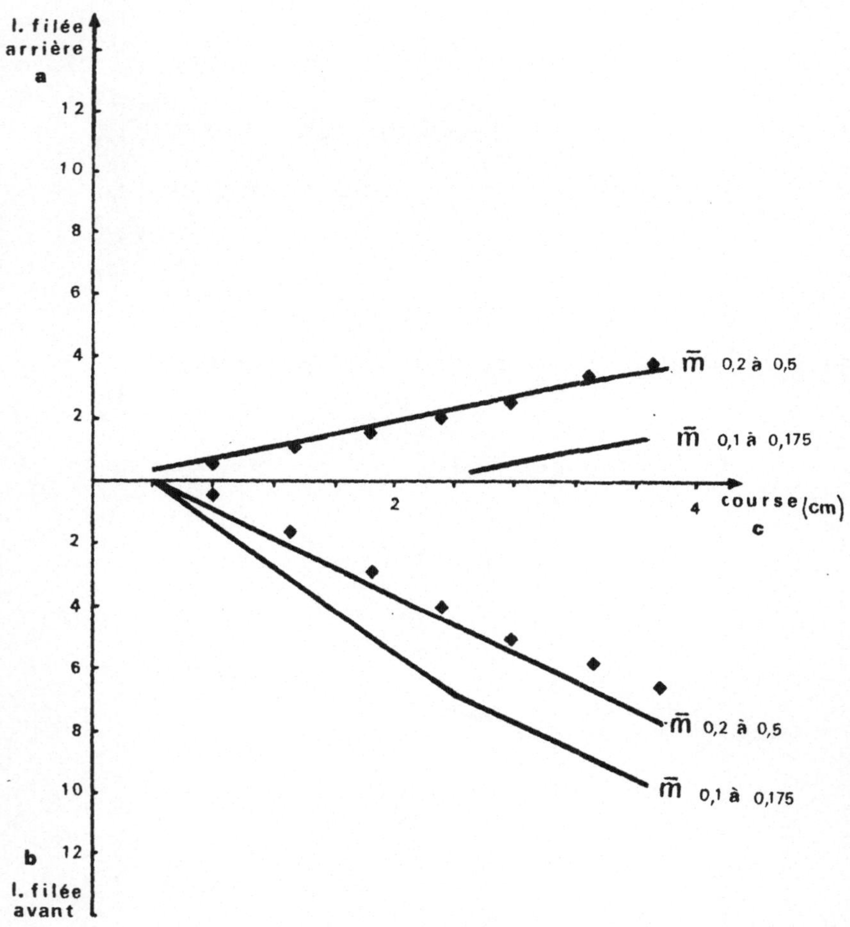

Figure 10 : Simultaneous can and rod extrusion ;

Extrusion force versus stroke

a	: Depth of the can	Diameter of the punch = 35 mm
b	: Length of the rod	Diameter of the die-insert = 23 mm
c	: Stroke	Cone angle (2 α) = 120°

◆ : experimental results

However in very few cases the energy difference between forward and backward extrusion during the stroke is too small to give a good prediction. This phenomenon of instability appears also during the production : very little change in the tool or in the lubrication can totally change the end shape of the component.

For an example, let us consider the influence of the friction on the end geometry of à component (fig. 10). In this case the prediction is not good until $\bar{m}$ has reached a limit value $\bar{m}_o = 0,2$.

Afterwards the friction coefficient $\bar{m}$ has no more influence. The difference between the two predicted geometries is quite important, the depth of the can can be twice smaller for $\bar{m} = 0,175$ as for $\bar{m} = 0,2$. But a difference $\Delta\bar{m} = 0,025$ has not any physical sense in the way that $\bar{m}$ is experimentally measured.

This example brings two conclusions :
- The simulation program and the physical reality are in accordance ; a
  very small difference on lubrication can in some caseshe the reason of
  a great difference on the end-geometry of the component. These cases are
  defined by a very small difference between the internal powers calculated
  in case of a pure rod or a pure can extrusion.
- The choice of the value of $\bar{m}$, which represents the ability of the
  interface-materials to be sheared, must be done with a great accuracy.
  The problem appears that the usual tests - as for instance the ring test -
  are perhaps not so good because the deformations are smaller than those
  which occur during the extrusions. These tests are further more not
  sensible enough.

6.2   Extrusion forces

On figure 11 a comparison between experimental results and the calculated datas is shown on an example. The general correlation diagram for the experimental plan is plotted on figure 12. We can see that the correlation is pretty good (5 to 10 %). The calculated values are always greater than the experimental ones. The reason is that the upper bound method has been used for the prediction.

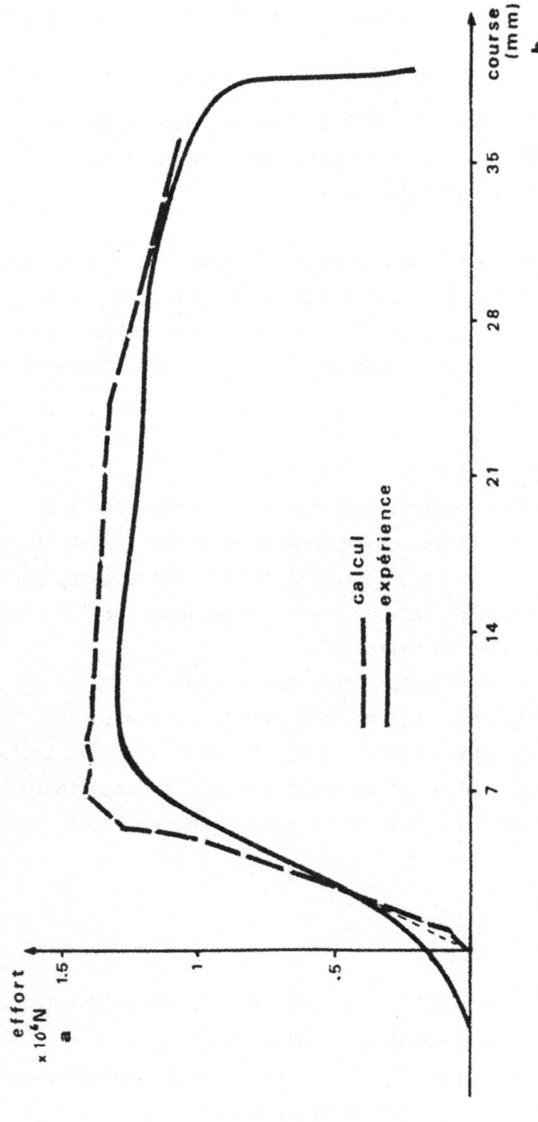

Figure 11 : Simultaneous can and rod extrusion ;
Extrusion force versus stroke

    a : Force   Diameter of the punch   = 35 mm

    b : Stroke   Diameter of the die-insert = 26,5 mm

                    Cone angle (2α)   = 90°

             : calculated results

             : experimental results

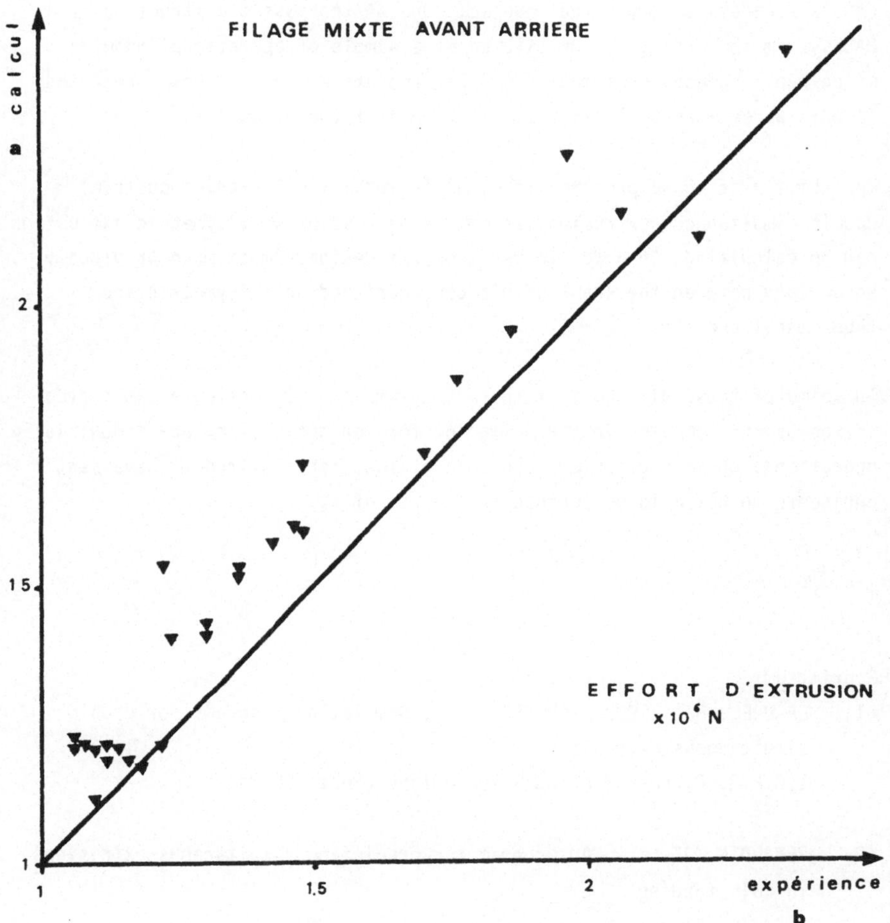

Figure 12 : Correlation diagram between :

. maximal calculated extrusion force (a), and

. maximal mesured extrusion force   (b)

7   Conclusion

CAD is nowadays a very actual subject. One of the possible structure of a
CAD system in forging is constituted of a sample of operational modules
which can be chosen on a menu (7). The programs which have been presented
in this paper were built in order to be one of these modules.

The ability of these programs to simulate automatically any industrial
double simultaneous extrusion is limited because only axisymetric extrusions
can be calculated. In order to use them the designer must make previously
an analogy between the shape of his component and an axisymetric and
theoretical one.

In spite of these difficulties these programs can be considered as a step
in the use of computer in the design of forging tools. They are industrially
operational on mini-computers and some of the french extruders have sent
engineers to CETIM to be learned in the use of it.

Schrifttum

(1)   BRAUDEL, HJ. ; RONDE-OUSTEAU, F. : Simulation Programs for double
       simultaneous extrusions.
       I.C.F.G. Plenary Meeting - Sunderland (Sept. 1978)

(2)   RENAUDIN, JF. : A CAD Program for predicting simultaneous extrusions
       in cold forging.
       I.C.F.G. Plenary Meeting - Lyngby (Sept. 1982)

(3)   BRAUDEL, HJ. ; RENAUDIN, JF. : Une méthode de simulation numérique
       du filage en forge à froid.
       Document CETIM - à paraître

(4)   RENAUDIN, JF. ; BRAUDEL, HJ. : Un modèle mécanique des filages
       combinés en forge à froid.
       Revue de Métallurgie - CIT (1983) Vol 80, 6, pp 501-509

(5)   RONDEPIERRE, JF. : Contribution à l'étude pratique et à la modélisa-
       tion numérique de l'extrusion des métaux.
       Thèse de Docteur-Ingénieur - Université de POITIERS - UER ENSMA

(6)   BRAUDEL, HJ. : Contribution à l'étude pratique et à la modélisation
       numérique des filages simultanés.
       Thèse de Docteur-Ingénieur - Ecole des Mines de PARIS

(7)   GEIGER, M. : Rechnerunterstützte Auslegung von Maschinen und
       Werkzeugen der Umformtechnik.
       Konstruktion 33 (1981) H1 pp 3 à 14 - H2 pp 49 à 53

(8)   GEIGER, R. : Kombinationen von Fliesspressverfahren.
       Draht 29 (1978) 5 - S. 223-229

Some Key Problems in the Tribology of Metalforming

John A. Schey, Department of Mechanical Engineering, University of Waterloo, Ontario / Canada

Summary

Over the last 20 years, knowledge pertaining to the interface between die and workpiece has expanded substantially, so that it is now legitimate to speak of the tribology of metal-forming. There are, nevertheless, many challenges remaining. Mixed-film lubrication is the prevalent mechanism when liquid lubricants are used, and the magnitude of friction and the roughness of the issuing surface are controlled, among others, by the proportion of the surface that is in boundary contact. Prediction of this proportion from basic process and lubricant parameters is still not possible. Interwoven with this problem is that of characterizing the roughness of the surface before, during, and after deformation, and of predicting the changes to be expected in the course of deformation. Metal-to-metal contact is often unavoidable, with undesirable consequences in terms of metal transfer, die pickup, and workpiece scoring. Prediction of adhesion tendencies from basic material proper-ties is still not possible, even though such prediction is necessary if die materials and coatings are to be chosen from first principles. Realistic modeling of the interface does not allow generalized simplifications as practiced when an average coefficient of friction or interface shear factor is used, and modeling of the interface with true regard for the active lub-rication mechanism remains an unsolved task for many situa-tions.

<u>List of Symbols</u>

b   Fraction of contact area in boundary contact
m   Frictional shear factor
p   Interface pressure
v   Sliding velocity
η Dynamic viscosity of lubricant
μ Coefficient of friction
σ_f Flow stress in uniaxial tension or compression
τ_f Flow stress in shear
τ_i Interface shear stress

0   Introduction

Considering the great antiquity of metalworking processes, it is somewhat surprising that the interface between die and workpiece has become only relatively recently an object of serious attention. Practitioners have, of necessity, recognized the importance of a good lubricant in reducing friction and wear, and in controlling surface finish. The interest of the theoretician was often limited to measurements of the interface shear strength or, rather, of the coefficient of friction, with little regard to the operative lubrication mechanism. Part of the reason for this neglect was the belief that, under the severe conditions prevailing in plastic deformation, only boundary lubrication can exist. Work on glass-lubricated hot extrusion and high-speed wire drawing pointed to the possibility of hydrodynamic lubrication; more detailed studies, conducted from the 50's onward on the effects of lubricant viscosity and process conditions on surface finish, indicated that mixed--film lubrication must be the dominant mechanism in many processes. As evidenced by the title of a monograph published in 1970 [1], wear was still of minor interest. Since then, more emphasis has been placed on a systematic consideration of the many material and process parameters that can have an effect, and one can now speak of the tribology of metalworking, with equal emphasis on friction, lubrication, and wear, as shown also by the title of a recent monograph [2].

Research in metalworking tribology tends to be fragmented and has been characterized by bursts of activity followed by rela-

tive lulls. The Institute für Umformtechnik of the University
of Stuttgart, under the leadership of Professor Kurt Lange, has
been one of the few institutions where research has had conti-
nuity, and it is a great pleasure indeed to present here some
thoughts on the state of the art and on some challenges of the
future.

1 Mechanisms of Lubrication

All lubrication mechanisms can be mapped on a modified and
expanded Stribeck diagram (Fig.1) [3]. It appears that most
non-liquid lubricants (including layer-lattice compounds, poly-
mer films, mono- or multimolecular boundary films and,
possibly, also reaction products formed with E.P. additives)
have a shear strength that increases roughly linearly with
interface pressure.

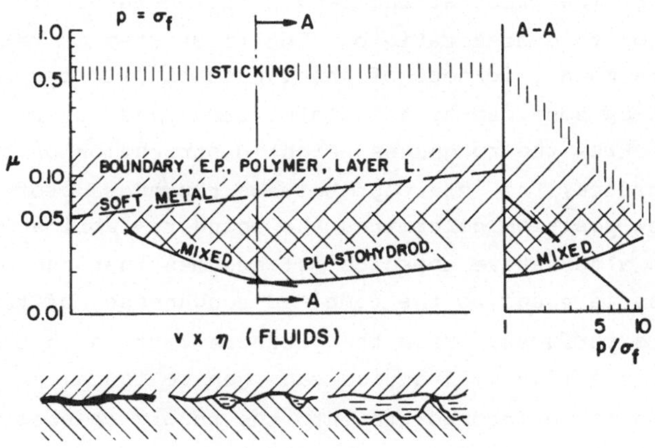

Fig. 1: Lubrication mechanisms in metalworking [5].

This results in a reasonably constant coefficient of friction,
on the order of 0.05 - 0.1, little affected by sliding speed v
or interface pressure p (expressed in Fig.1 as a multiple of
the uniaxial flow stress σ_f of the workpiece material). Soft
metals are usually strain-rate sensitive at the temperature of
their application, hence μ rises slightly with sliding velocity.

Complications begin when the lubricant is liquid. Theories of plastohydrodynamic lubrication [4] are reasonably successful in predicting film thicknesses, but results of rolling experiments with negative forward slip indicate that calculated tractions may be in error [5]. A fully valid treatment is still not available and would have to incorporate proper allowance for heating and also for roughening of the workpiece surface.

At the usual speeds and with most liquid lubricants, full separation of the surfaces is unlikely and mixed-film lubrication prevails. Some of the lubricant is entrapped to form hydrostatic/hydrodynamic pockets of very low shear strength (or μ_h), while some considerable portion (b) of the surface is in boundary contact. The coefficient of friction drops with decreasing boundary-lubricated contact area

$$\mu_{mix} = b\mu_b + (1 - b)\mu_h$$

and one of the greatest challenges facing theory is the prediction of this area ratio b. The first step is relatively easy: the mean thickness of the lubricant film carried into the deformation zone can be calculated from plastohydrodynamic theory. From the roughness height distribution of the workpiece surface, the intercept between roughness profile and the interface position determines the boundary-lubricated area [6]. An alternative approach [4] assumes that the mean film thickness is equal to the combined roughnesses of the die and workpiece surfaces. From the pressure carried in the lubricant pockets, the (1 - b) fraction can then be found. A remaining challenge is to incorporate into the theory changes in surface roughness occurring during the deformation process itself.

2 Effects of Surface Roughness

The above view regards the fluid entrapped into hydrodynamic pockets as essentially immobile, expanding with the surface, but not participating in the lubrication of the boundary contact zone. However, upsetting experiments with cylinders of grooved end faces showed that the volume of grooves gradually diminishes in the course of compression, while the ridges gradually roughen (Fig.2) [7]. This indicates that the trapped

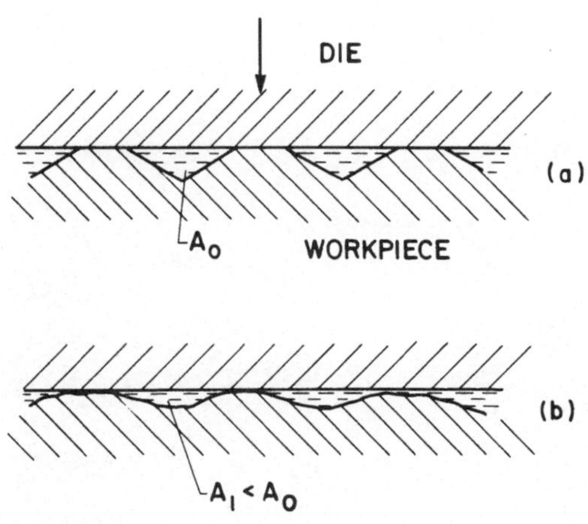

Fig. 2: Deformation of grooved end face in
upsetting a cylinder with a viscous lubricant [7].

fluid is a source of lubricant when external supply is impossible or insufficient in quantity. It also explains the great success of electrolytic zinc or tin coatings in preventing lubricant breakdown in severe operations such as wire drawing or ironing. The microscopic channels present in the as-deposited coatings fulfill the role of lubricant reservoirs, preventing damage to the surface (Fig.3.a), while a smooth, hot-dipped coating suffers much damage (Fig.3.b) [8].

In general, the surface roughnesses of die and workpiece play decisive roles in the success of lubrication [2,9]. A complicating factor is that the surface microgeometry changes as a result of the lubricated deformation process itself [9,10]. Hence, the two most significant remaining challenges are a tribologically meaningful characterization of surface roughness, and a systematic, quantitative treatment of the interactions of roughness and changes in roughness with tribological processes.

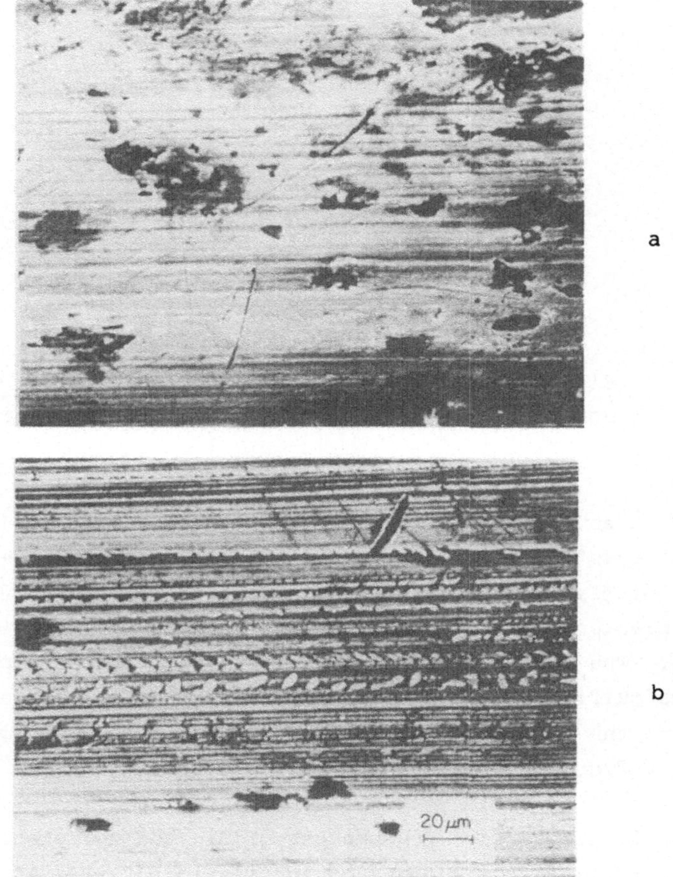

Fig. 3: SEM photographs of high-strength, patented, 0.8% carbon-steel wire drawn through 17 holes with calcium soap (a) electrolytic zinc coating, (6) hot-dipped zinc coating [7].

3 Metal-to-Metal Contact

Contact between workpiece and die material have a dominant
effect on the success of a process. In the extreme, no lubricant is interposed between the two materials and the practi-
cally and theoretically important condition of sticking friction is attained. The treatment of this situation presents
more problems than usually suspected, even if there is no
actual sticking (adhesion) between die and workpiece. In
upsetting a cylinder, the accompanying inhomogeneous
deformation keeps interface pressures low, barely above the
uniaxial flow stress, hence pressures calculated from the slab
theory for homogenous compression are much too high [11]. A
similar situation appears to hold for ring compression [12].
Preliminary results of experiments with grooved anvils show
that the pressure-multiplying factor is substantially lower
than predicted from upper-bound theory (Fig.4). This could be
simply the result of

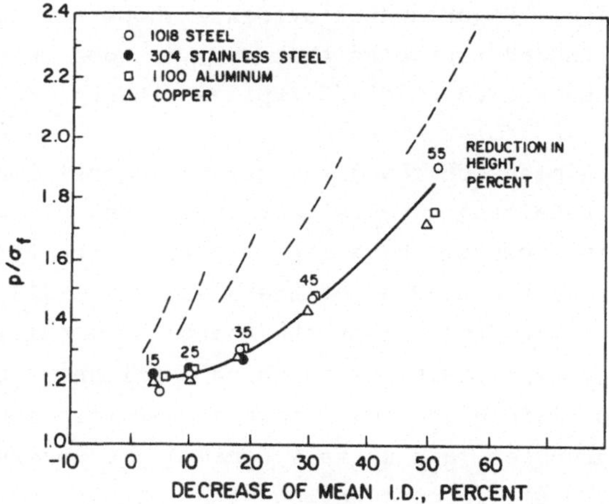

Fig. 4: Pressure increase in compressing rings of
6:3:2 aspect ratio with sticking friction (grooved
anvils) (broken lines: theory) [12].

non-homogeneous deformation, with barreling contributing to a
lowering of interface pressures. A further cause could be

strain hardening of the intensively sheared material adjacent
to the die surface. This should, however, simply shift the
zone of shearing further into the body of the workpiece, and
the effective shear strength should still be $0.5\sigma_f$ or
$0.577\sigma_f$. However, further results, not shown here, indicate
that material properties are a source of variation too, particularly with two-phase materials such as brasses and heat-
treated aluminum alloys.

Somewhat paradoxically, adhesion and thus the danger of metal
transfer (die pickup) is relatively small with sticking friction, because contaminant films are capable of preventing direct metallic contact as long as there is no relative sliding.
In contrast, die pickup is of major concern in mixed-film
lubrication and, in the event of film breakdown, also in hydrodynamic lubrication. Die pickup is controlled by adhesion
between workpiece and die, and an ideal die would not allow
metal transfer even in the absence of a protective contaminant
or lubricant film. Experimental evidence shows that adhesion
and pickup are indeed variables that can, to some extent, be
controlled by the choice of die/workpiece material combination.

Coatings offer a means of changing surface composition and
hardness while retaining the desirable toughness of the sub-
strate, and some successes have been reported even for such
difficult tasks as the drawing of stainless steel [13,14].
There is, however, a danger that the changed composition of the
surface might prevent reactions on which lubricant action
relies. This is visible in Fig.5 where higher friction and
greater metal transfer (and surface damage) are observed on
using a boronized die in conjunction with stearic acid;
obviously, beneficial reaction was prevented by the die coating
[15].

Despite the great advances made in understanding adhesion [16],
there are still no fundamental guidelines that would allow the
choice of die material from first principles. Continuing
experimental work in several centers has succeeded in

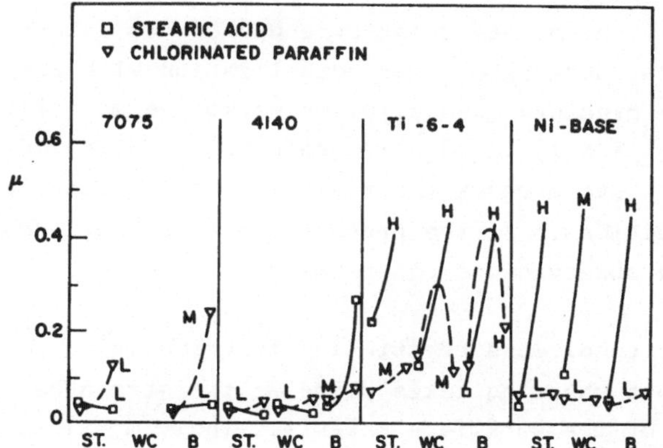

Fig. 5: Effects of anvil material (ST=tool steel; WC = tungsten carbide; B = boronized steel) on coefficient of friction and surface damage (L = light; M = medium; H = heavy) in twist compression [15].

elucidating many factors that affect lubricating mechanisms, film breakdown, and die pickup [17-19]. What is still missing is a working hypothesis for predicting the consequences of metal-to-metal contact.

Adhesion is a significant factor also in determining wear in a given system, not only because adhesive wear is often encountered, but also because metal transfer can lead to accelerated abrasive wear and to scoring damage of the workpiece surface. For these reasons too, prediction of adhesion is highly desirable.

4 Modeling of the Interface

From the practical point of view, the success of lubrication can be readily judged from the absence of die pickup and tool wear, from the generation of an acceptable surface finish and, in general, trouble-free operation. Nevertheless, there is

still a need for a realistic mathematical modeling of the
interface so that valid process models can be set up.
Obviously, the actual shear strength of the interface τ_i is
the closest descriptor, but normalization with respect to
interface pressure ($\mu = \tau_i/p$) or workpiece material flow
stress ($\tau_i = m\sigma_f$) is usually preferable. Such modeling
must take into account the real response of lubricants to process conditions. At the present state of knowledge, the
following observations can be made:

a. Under conditions of sticking friction it is reasonable to
assume that shearing takes place at the strain-hardened shear
strength of the workpiece material, thus, $m = 1$ and μ is
inapplicable. This comforting situation may change as more is
learnt about deformation with sticking friction.

b. Soft metal coatings have a largely pressure-independent
shear strength which is, however, unlikely to retain a fixed
relationship to σ_f. Thus, neither μ nor m are likely to suf-
fice as descriptors. The liquid or semi-solid lubricant normally superimposed on a metal coating further complicates the
situation.

c. Polymers, layer-lattice compounds, boundary (and possibly
E.P.) films have a pressure-dependent shear strength. Thus,
m is inappropriate and a constant μ gives a reasonable representation.

d. Liquid-lubricated interfaces operating in the mixed-film
regime cannot, obviously, be described by m. The problem is
that μ also varies with interface pressure, lubricant viscosity, sliding speed, and even workpiece material [18,19]. Until
a complete theory of mixed-film lubrication becomes available,
an estimate of a variable μ is perhaps the only reasonably
accurate approach.

e. In plastohydrodynamic lubrication the interface is best
described by μ, although further advances will be needed

before tractions can be adequately predicted [19]. Fortunately, interface pressures and forces are so close to the frictionless values that even a large error in tractions has little disturbing effect.

Evidently, no single approach to modeling the interface can possibly describe the great variety of situations encountered, and future efforts could most profitably be directed towards predicting τ_i from basic principles.

References

[1]   Schey, J.A. (ed.): Metal Deformation Processes: Friction and Lubrication, Dekker, New York (1970).

[2]   Schey, J.A.: Tribology in Metalworking: Friction, Lubrication, and Wear, American Society for Metals, Metals Park, OH (1983).

[3]   Schey, J.A.: Modeling of the Tool-Workpiece Interface, in Lippmann, H. (ed.): Metal Forming Plasticity, Springer, Berlin (1979), pp. 336-348.

[4]   Wilson, W.R.D.: Friction and Lubrication in Sheet Metal Forming, in Koistinen, D.P. and Wang, N.M. (ed.): Mechanics of Sheet Metal Forming, Plenum, New York (1978), pp. 157-177.

[5]   Reid, J.V.; Schey, J.A.: Full Fluid Film Lubrication in Aluminum Strip Rolling, ASLE Trans. 21 (1978) 3, 191-200.

[6]   Tsao, Y.H.; Sargent, L.B.: A Mixed Lubrication Model for Cold Rolling of Metals, ASLE Trans. 20 (1977) 1, 55-63.

[7]   Hoba, R.: Lubricant Entrapment in Upsetting, Project Report, University of Waterloo, 1981.

[8]   Apel, G.; Nünninghoff, R.: Veränderung der Zinkschicht beim Ziehen verzinkter dünner, hochfester Stahldrähte, Stahl u. Eisen 100 (1982) 21, 1247-1253.

[9]   Schey, J.A.: Surface Roughness Effects in Metalworking Lubrication, Lubric. Eng. 39 (1983) 6, 376-382.

[10] Dannenmann, E.: Oberflächen- und Randzonenbeeinflussung durch Umformen und Schneiden, Techn. Mitt. 73 (1980) 11/12, 893-901.

[11] Schey, J.A.; Venner, T.R.; Takomana, S.L.: The Effect of
     Friction on Pressure in Upsetting at Low Diameter-to-
     Height Ratios, J. Mech. Work Tech. 6 (1982) 1, 23-33.

[12] Bingham, I.: Material Effects in Ring Compression, Project
     Report, University of Waterloo, 1983.

[13] Schlosser, D.: Verwendung oberflächenbehandelter Werkzeuge
     beim Blechumformen, Ind.-Anz. 98 (1976) 6, 158-162.

[14] Kerspe, J.H.: Lösung tribologischer Probleme beim
     Abstreckgleitziehen des nichtrostenden austenitischen
     Werkstoffs 1.4301, Ind.-Anz. 102 (1980) 91, 60-62.

[15] Schey, J.A.; Newnham, J.A.: Effect of Die Composition on
     Lubrication in Metalworking, Lubric. Eng. 26 (1970) 4,
     129-137.

[16] Buckley, D.H.: Surface Effects in Adhesion, Friction,
     Wear, and Lubrication, Elsevier, Amsterdam (1981).

[17] Kawai, N.; Nakamura, T.; Dohda, K.: Anti-Weldability Test
     in Metal Forming by Means of Strip-Ironing Type Friction
     Testing Machine, Trans. ASME, Ser.B, J. Eng. Ind. 104
     (1982) 4, 375-382.

[18] Gräbener, T.; Lange, K.: Tribological Conditions in Cold
     Metal Forming, in Proc. 10th NAMRC, Society of
     Manufacturing Engineers, Dearborn, MI (1982), pp. 122-129.

[19] Kudo, H.; Tsubouchi, M.; Takada, H.; Okamura, K.: An
     Investigation into Plasto-Hydrodynamic Lubrication with a
     Cold Sheet Drawing Test, CIRP Annals, 31 (1982) 1,
     175-180.

Optimale Reibbedingungen beim geschmierten Tiefziehprozess

Josef Reissner, Institut für Umformtechnik, ETH-Zürich

1 Einleitung

Beim Tiefziehen bestimmen die Eigenschaften des Ziehteilwerkstoffes und des Schmierstoffes sowie die Werkzeuggeometrie und die Umformparameter die Verfahrensgrenzen: das Auftreten von Rissen und die Bildung von Falten. Beide Versagensarten werden durch die Reibvorgänge in so grossem Masse beeinflusst, dass für eine zuverlässige Berechnung des Arbeitsbereiches möglichst genaue Reibungszahlen notwendig sind. Das Ziel der vorliegenden Arbeit soll darin bestehen, den Reibeinfluss auf die Faltenbildung im Flansch und auf die Entstehung der örtlichen Einschnürung im Bereich der Stempelrundung zu untersuchen. Gleichzeitig soll versucht werden, die Reibparameter in den einzelnen Kontaktzonen zwischen Werkzeug und Werkstück festzulegen und zu quantifizieren. Unter Berücksichtigung dieser Einflussgrössen wird ein Reibmessgerät zur Ermittlung von μ-Werten vorgestellt.

2 Faltenbildung und Niederhalterreibung

Beim Tiefziehen erfolgt die Umformung im Flanschbereich unter radialen Zug- und tangentialen Druckspannungen. Dieser Spannungszustand kann zur Bildung von Falten führen, wenn das Stempeldurchmesser-Blechdicken-Verhältnis und das Ziehverhältnis gross sind. Zur Berechnung des Auffaltvorganges wird als Modell eine kreisförmige Scheibe verwendet, die am Innenradius gelagert ist und durch radiale Zugspannungen beansprucht wird. Die daraus resultierenden Druckspannungen bewirken innerhalb der Flanschbreite das Ausbrechen aus dem ebenen in den geknickten Zustand.

Die Lösung des plastischen Knickvorganges erfolgt durch Anwendung des Minimalprinzipes von Rayleigh [1,2]. Der wichtigste Berechnungsschritt ist die Verknüpfung der Formänderungen mit den Spannungen bzw. den Krümmungen mit den Momenten zur

Ermittlung der Energieumsetzungen zwischen dem ebenen und gefalteten Flansch (Bild 1).

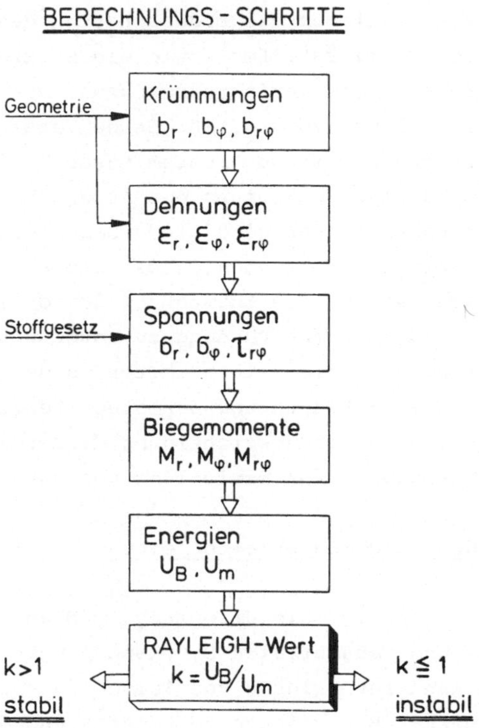

BERECHNUNGS - SCHRITTE

Geometrie →

Krümmungen
b_r , b_φ, $b_{r\varphi}$

Dehnungen
ε_r , ε_φ, $\varepsilon_{r\varphi}$

Stoffgesetz →

Spannungen
σ_r , σ_φ , $\tau_{r\varphi}$

Biegemomente
M_r , M_φ, $M_{r\varphi}$

Energien
U_B , U_m

RAYLEIGH-Wert
$k = U_B / U_m$

$k > 1$ ⇐ ⇒ $k \leqq 1$
stabil   instabil

Bild 1: Schritte für die k-Wert-Berechnung

Dabei ist die Biegeenergie (U_B) der Energieanteil, der aufgewendet wird, um den gekrümmten Zustand der Platte zu erhalten. Gleichzeitig wird durch diesen Vorgang der Flansch entspannt, was mit einer Energiefreisetzung (U_m) verbunden ist.

Ein Kriterium für das Stabilitätsverhalten stellt der Quotient

dieser Energieanteile, der sogenannte Rayleigh-Wert k dar. Ist
k > 1, so sind die während der Umformung im Flansch auftretenden Spannungen unterkritisch, d.h. der Flansch bleibt eben.
Eine Bildung von Falten ist dann zu erwarten, wenn mit k ≤ 1
ein kritischer bzw. überkritischer Zustand vorliegt. Nach dem
energetischen Minimalprinzip nimmt der Flansch denjenigen geknickten Zustand an, bei dem die Potentialdifferenz zum ebenen
stabilen Grundzustand minimal ist. Dies bedeutet, dass sich im
Flansch diejenige Anzahl von Falten einstellen wird, bei der
der k-Wert sein Minimum aufweist.

Für die praktische Anwendung ist es zweckmässig, den bezogenen
k-Wert in Abhängigkeit vom ρ-Wert darzustellen (Bild 2).

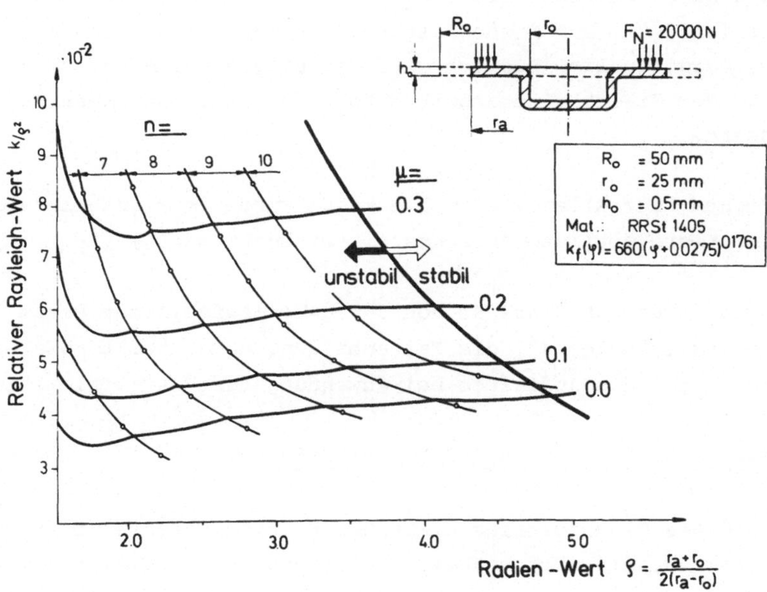

Bild 2: Einfluss des μ-Wertes auf das Stabilitätsverhalten

In diesem Diagramm wird der Verlauf der k = 1-Werte als Stabi-
litätskurve und die Verbindung aller minimalen bezogenen k-Wer-
te als Faltenzahlkurve bezeichnet. Der Stempelweg wird mit dem
ρ-Wert angegeben, der das Verhältnis des mittleren Radius zur
Flanschbreite charakterisiert. Die für die Untersuchung des
Stabilitätsverhalten notwendige Spannungs- und Formänderungs-
verteilung wurde mit Hilfe der finiten Differenzenmethode unter
Berücksichtigung des Niederhalters berechnet [3,4]. Bekannt-
lich verändern die vom Niederhalter herrührenden Reibkraftkom-
ponenten den inneren Zustand. So ergeben sich bei konstanter
Niederhalterkraft und grösserem μ-Wert nicht nur eine geringere
Dickenzunahme sondern auch grössere radiale Zugspannungen, die
wiederum kleinere tangentiale Druckspannungen zur Folge haben.
Das gilt auch, wenn bei einer konstanten Reibungszahl die Niederhalterkraft zunimmt. Während der Dickeneinfluss den k-Wert
verkleinert, erfolgt durch die Aenderung des Spannungszustandes
eine k-Wert Erhöhung, die den Einfluss der Dicke überkompen-
siert (Bild 2). Der Niederhalter wirkt zusätzlich auch als äus-
sere Energiequelle. Dadurch wird der Biegeenergieterm um den
Betrag der Niederhalterarbeit erhöht und damit der k-Wert vergrössert.

Mit einem speziellen Tiefziehwerkzeug wurde im Flanschbereich
der mittlere örtliche Reibungsbeiwert bestimmt [5].

In Bild 3 ist der Einfluss des μ-Weg-Verlaufes einer Paste
und eines Ziehöles auf die Faltenbildung dargestellt. Wie erwartet, wird das Auffalten bei Anwendung der Paste erleichtert.

3 Makroschmiertasche und Schmierstoffdruck

Während des Ziehprozesses nimmt der Flanschquerschnitt eine
Keilform an, die sich ständig ändert und ein partielles Auf-
liegen des Niederhalters verursacht. Aufgrund der tangentialen
Druckspannungen verdickt sich der Flanschrand, während die radialen Zugspannungen in Ziehringrundungsnähe für eine Abnahme
der Blechdicke verantwortlich sind. Mit zunehmender Ziehtiefe
wird sowohl der Bereich der Verdickung als auch die Verdickung

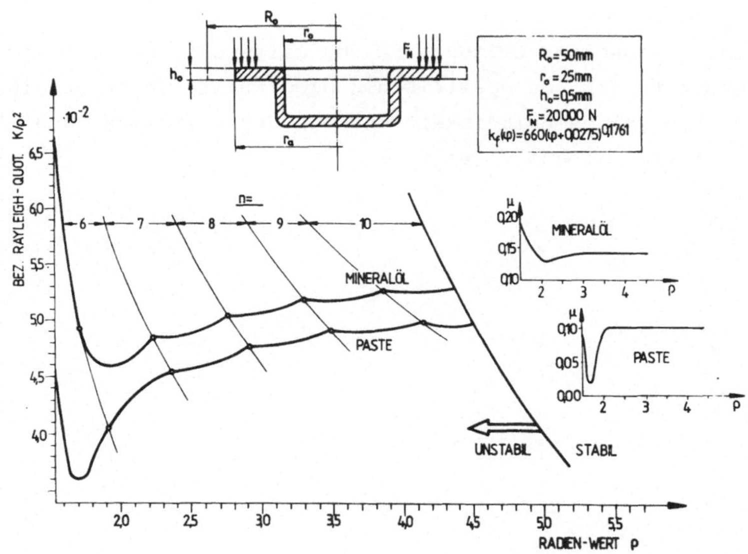

Bild 3: Einfluss des µ-Weg-Verlaufes von Schmierstoffen
auf die Faltenbildung

selbst vergrössert. Der Dickenverlauf in Verbindung mit den
Flächenpressungen, die auf der Makrokontaktfläche und in der
Ziehringrundung wirksam werden, können zur Bildung einer Makro-
schmiertasche führen [6].

Sind die geometrischen Verhältnisse, die Umformparameter und
die Schmierstoffdaten bekannt, so lassen sich der Druck und die
Schichtdicke des Schmierstoffes mit Hilfe der Strömungsmechanik
berechnen [7,8]. Bereits beim Aufbringen der Schmierstoffschicht
ist zu achten, dass die Schichtdicke unter Niederhalterbelastung
mit derjenigen Spaltdicke übereinstimmt, die sich beim Ziehen
einstellt. Dabei gelten folgende Gesetzmässigkeiten:
1. Die Schichtdicke des Schmierstoffes (h_N.. niederhalterseitig,
h_Z.. ziehringseitig) nimmt umgekehrt proportional zur Wurzel

aus der Zeit t ab.

2. Die Schichtdickenänderung $\frac{dh_N}{dt}$ ist direkt proportional zu $\frac{FN}{\eta}$ und zur Kubik der jeweiligen Schichtdicke h_N und h_Z.

Wie die Berechnungen zeigen, ist der Zeitpunkt, bei dem der Schmierstoffdruck den spezifischen Niederhalterdruck erreicht, von der Stempelgeschwindigkeit und von der Viskosität des Schmierstoffes abhängig (Bild 4).

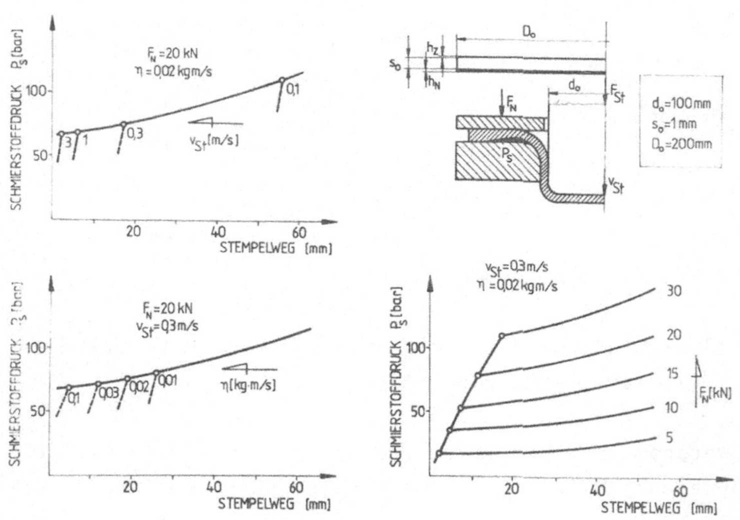

Bild 4: Einfluss der Stempelgeschwindigkeit v_{St}, der Viskosität η und der Niederhalterkraft F_N auf den Schmierstoffdruck

Die Einstellung erfolgt umso früher, d.h. bei geringerer Ziehtiefe, je grösser die Viskosität und die Stempelgeschwindigkeit und je kleiner die Niederhalterkraft ist. Die Höhe des Druckes, die für die Verhinderung der Falten im nichtunterstützten Flanschbereich verantwortlich ist, wird durch die Grösse der Niederhalterkraft bestimmt.

4 Stempelreibung und Bodenreisser

Die zum Umformen notwendige Ziehkraft wird vom Stempel in den
Ziehteilboden eingeleitet und in den Flansch übertragen. Die
grösste Kraft wird dann in die Umformzone übertragen, wenn die
kritische Zone nach einer Zug-Druck-Umformung am Auslauf der
Stempelkantenrundung zu liegen kommt. Die zum Bruch führende
örtliche Einschnürung wird nicht nur vom Spannungszustand bestimmt, sondern hängt auch vom Formänderungsweg ab. Eine gute
Schmierung unter dem Niederhalter und in der Ziehringrundung
verbessert die Ziehergebnisse, eine geringe Stempelreibung be-
wirkt das Gegenteil.

Für die Berechnung der Verfahrensgrenze geht man davon aus,dass
der Bereich der diffusen Einschnürung noch voll ausgenützt werden kann. Als Versagenszeitpunkt wird der Moment betrachtet, in
dem das Material durch Bildung einer im Neigungswinkel von der
Anisotropie abhängigen Bruchfläche innerhalb der Einschnürzone
versagt. Dabei kommt es zum Abgleiten über die Blechdicke. Als
Ausgangspunkt für die Theorie dient der für diese Versagen
charakteristische Rückfall in den elastischen Werkstoffzustand:
der Rückfall jener Bereiche, die die kritische Zone umgeben.
Die Unterscheidung der beiden Zustände erfolgt dabei durch den
Vergleich der im elastischen Stoffzustand sich einstellenden
und im plastischen Zustand effektiv vorhandenen Spannungsfelder.

Durch Definition einer die Abweichung in der Spannungsverteilung charakterisierenden Funktion ergibt sich eine Möglichkeit,
innerhalb der Einschnürzone die rissgefährdete Lage zu ermitteln. Der Riss entsteht bei duktilen Werkstoffen in dem Moment,
in dem diese Richtung mit der vom Spannungszustand und vom
R-Wert abhängigen Richtung der reinen Scherung zusammenfällt[9].

Will man den Ort und den Zeitpunkt des Versagens feststellen,
so muss bei Auftreten der ersten Einschnürung (strain-propagat-
ion-instability) von der zweiachsigen Spannungsbetrachtung zu
einer dreiachsigen übergegangen werden (Bild 5). Die Spannungs-
berechnung für unregelmässige Ziehteile ist nur mit der Methode

der finiten Elemente durchführbar.

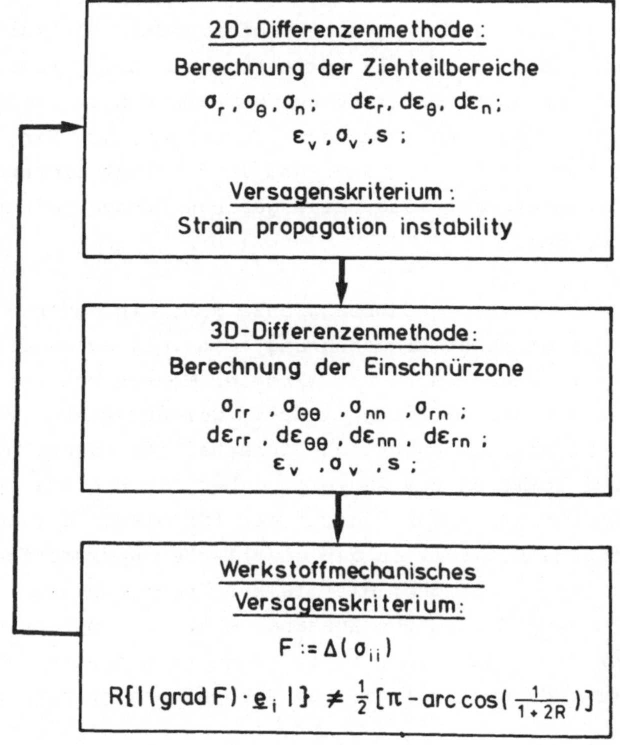

Bild 5: Berechnung von rotationssymmetr.Ziehteilen

Hingegen kann die Spannungsverteilung von rotationssymmetrischen Teilen entweder durch geeignete Ansatzfunktionen oder
durch Erweiterung des 2D-Differenzenverfahrens auf eine 3D-
Berechnung ermittelt werden. Durch Anwendung des werkstoffme-
chanischen Versagenskriteriums ist es möglich, Lage und Zeitpunkt des Risses zu bestimmen. In guter Uebereinstimmung mit
den Versuchen zeigen die ersten Berechnungsergebnisse, dass
ein Zusammenhang zwischen dem Grenzziehverhältnis und den Reib-

verhältnissen vorallem im Stempelrundungsbereich besteht
(Bild 6).

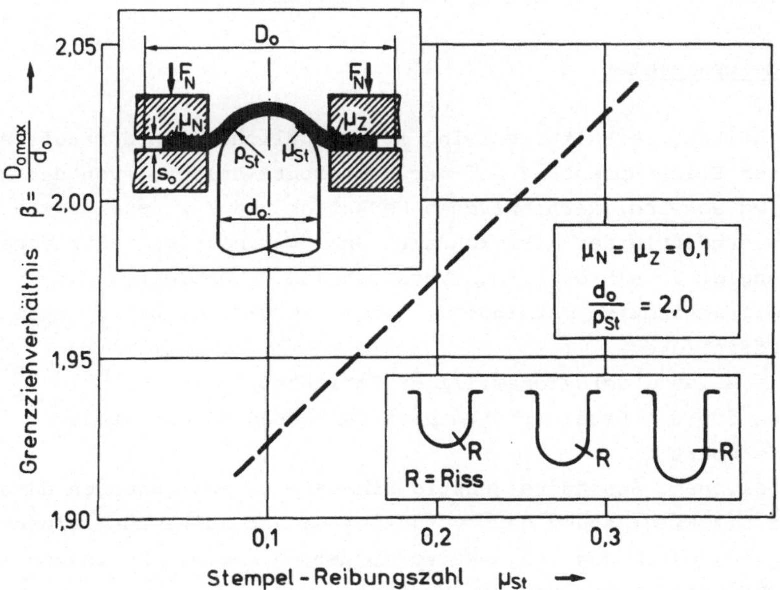

Bild 6: Einfluss der Stempelreibung auf die Lage der Risse
und auf das Grenzziehverhältnis

Ein Verringern des Reibbeiwertes verschiebt die höchstbean-
spruchten Stellen in Richtung Stempelmitte zu kleineren Durch-
messern und damit zu kleineren tragbaren Querschnitten. Eine
Verbesserung des Ziehverhältnisses kann dadurch erzielt werden,
dass ein vom Verhältnis $\dfrac{d_o}{\rho_{St}}$ abhängiger Minimal-μ-Wert eingestellt wird.Dadurch wird sich der Ort des Versagens zum Aus-
lauf der Stempelkantenrundung hin verschieben.

Aber nicht nur eine grosse Stempelreibung, sondern auch ein
hoher R-Wert, niedrige Reibungsverluste zwischen Werkzeug und

Werkstück im Bereich des Flansches und der Ziehringrundung sowie kleine Verluste infolge Biegung begünstigen die Verlagerung
der kritischen Zone. Die besten Ergebnisse werden dann erreicht,
wenn die kritische Zone nach einer Druck-Zug-Vorverformung am
Auslauf der Stempelrundung zu liegen kommt.

5. Reibparameter

Die Reibungsverhältnisse sind im wesentlichen von der auftre-
tenten Flächenpressung auf der Makrokontaktfläche, von der Relativgeschwindigkeit zwischen Werkstück und Werkzeug, sowie
vom Verhältnis der Mikrokontaktfläche zur Makrokontaktfläche
abhängig. Um ein örtliches Versagen und Durchbrechen der
Schmierschicht zu verhindern, ist es notwendig, Bedingungen zu
schaffen, die
- die Bildung adhäsionswilliger Schichten
- den für die Regenerierung erforderlichen Schmierstoff-
  transport
ermöglichen. Besonders günstig sind die Reibbedingungen dann,
wenn die Einglättung nach vorausgegangener Aufrauhung einsetzt.
Für die Bestimmung der μ-Werte in den einzelnen Reibzonen ist
die Kenntnis der Parameter Voraussetzung.

Als Folge der Niederhalterpressung wird sich trotz unterschiedlicher Dickenformänderung im Flansch eine Makrokontaktfläche
mit konstanter Blechdicke einstellen. Eine Zunahme des Anteils
der Mikrokontaktfläche erfolgt sowohl in Richtung des Flansch-
randes als auch mit steigender Ziehtiefe. Bei Werkstoffen mit
Flächenanisotropie verdickt sich der Flansch bevorzugt in den
Richtungen niedriger R-Werte, sodass die reale Mikrokontaktfläche hier wesentlich grösser ist. Während des Tiefziehpro-
zesses nimmt die Flächenpressung und die Gleitgeschwindigkeit
zu.

In der Ziehringrundung tritt eine wesentlich höhere Flächenpressung auf und die Relativgeschwindigkeit entspricht der
Stempelgeschwindigkeit. Trotz der vergrösserten Flächenpressung
und der damit verbundenen Einebnung der stark durch freie Um-

formung aufgerauhten Oberfläche, sind immer noch Mikrotäler
vorhanden, in denen sich hydrostatische Drücke ausbilden können und die die Grenzreibflächen mit Schmierstoff versorgen.
Bei Schmierstoffen mit höherer Viskosität liegen sie in grös-
serer Anzahl vor.

Im Stempelrundungsbereich wird die Innenseite infolge Mikro-
prägevorgänge zwischen Blechoberfläche und Stempelrundung ge-
glättet. Die Relativgeschwindigkeit ist in diesem Reibbereich
klein und die Flächenpressung liegt noch höher als in der Ziehringrundung.

6   Reibmessgerät

Von den vielen verfahrensfreien Modellversuchen hat der Strei-
fenziehversuch mit Umlenkung, der die Beanspruchungen beim Tiefziehen am besten simuliert, die grösste Bedeutung erlangt [6].
Bei diesem Versuch wird ein Streifen bei einer Umlenkung von
90° jeweils über einen drehbar gelagerten und über einen festen
Formkopf gezogen (Bild 7). Mit einem geringen Versuchsaufwand
ist es möglich, den Einfluss unterschiedlicher Reibparameter
auf die Reibverhältnisse zu untersuchen.

Von grösster Wichtigkeit ist die Relativgeschwindigkeit zwischen Werkzeug und Werkstück. Dabei gilt es zu beachten, dass
bei höheren Geschwindigkeiten Probleme mit der thermischen
Stabilität des Schmierstoffes auftreten. Um den Geschwindig-
keitsbereich von 0-5m/s abfahren zu können, ist die Probe in
einem geschlossenen Seilzug integriert. Der Vorteil liegt darin, dass die kleinen Massen grosse Beschleunigungen zulassen
und bereits nach einem kurzen Anfahrweg die Prüfgeschwindigkeit
erreicht wird. Die Proben können unverformt oder mit Hilfe des
Seilspanngetriebes verformt bis zu einem bestimmten Dehnwert
gezogen werden. Die plastische Dehnung wird auf einfachste
Weise durch Aenderung der Seilgeometrie erzeugt. Durch die Verwendung von Umsteckrollen ist es möglich, dass bei jedem Dehnwert die gesamte Probenlänge als Prüfstrecke zur Verfügung
steht.

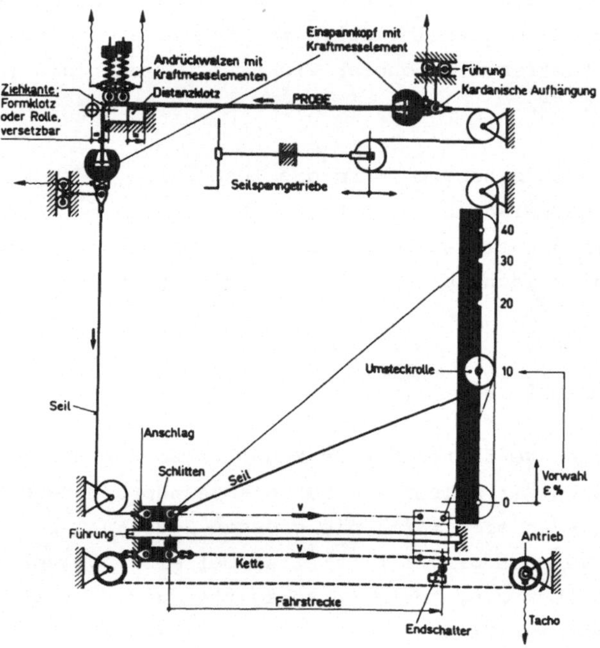

Bild 7: Reibmessgerät

Neben der Prüfgeschwindigkeit spielt die Flächenpressung eine
wesentliche Rolle. Deshalb ist es wichtig, dass die Blechprobe
unter annähernd konstanter Flächenpressung gleichmässig am
Formkopf anliegt. Die spezifische Flächenpressung kann in diesem Fall relativ genau bestimmt werden.

Der dritte Parameter, nämlich Makro-Mikro-Kontaktflächen-Ver-
hältnis wird durch Vorverformung mit Hilfe von Walzen eingestellt. Damit simuliert man auf der Blechoberfläche das Aufein-
anderfolgen der Aufrauhung und der Einebnung, und zwar unter
Bedingungen, wie sie beim Blechumformen auftreten. Die ersten
Messungen mit diesem neu entwickelten Reibmessgerät haben ergeben, dass dieser Einfluss berücksichtigt werden muss.

7 Summary

The influence of friction on the formation of folds under the
blankholder, has been investigated, based on the finite
differential method and the Raleigh minimal principle. A calculation based on the fluid mechanics principle has been used
to quantify the influence of various parameters on the level
of the pressure of the lubricating agent for the zones, which
were unprotected.

The location of the crack, which determines the magnitude
of the limiting drawability, has been decided in accordance
with a new failure criterion for materials, combined with the
differential shell calculations. Finally a strip-drawing
equipment has been introduced, through which the frictional
effects in individual zones could be simulated and the value
of μ as a function of the important friction parameters could
be calculated.

Schrifttum

[1] Meier, M.; Reissner, J.: Bildung und Verhinderung von Fal-
     ten im Flansch eines Tiefziehteiles.
     VDI-Berichte Nr. 45o, 1982, S.173-179

[2] Meier, M.; Reissner, J.: Instability of the Annular Ring
     as Deep-Drawn Flange under Real Conditions.
     Annals of the CIRP 32 (1983) 1, S.187-19o

[3] Woo, D.M.: Analysis of the Cup Drawing Process.
     J.Mechan.Eng.Sc. 6 (1964), 116

[4] Reissner, J.; Hora, P.: Hydromechanical Deep-Drawing.
     Annals of the CIRP 3o (1981) 2, S.2o7-21o

[5] Kasik, N.; Reissner, J.: Reibungsverhältnisse im Flansch
     beim Tiefziehen rotationssymmetrischer Teile.
     Blech Rohre Profile 27 (198o) 5, S. 3o3-3o8

[6] Witthueser, K.P.: Untersuchung von Prüfverfahren zur Be-
     urteilung der Reibverhältnisse beim Tiefziehen.
     Diss. TU Hannover 198o

[7] Reissner, J.; Stuhlträger, E.: Berechnung des hydrosta-
     tischen Druckes in einer Makroschmiertasche.
     IU-ETH-Zürich 1982

[8] Reissner, J.; Werner, A.: Computerprogramm zur Berechnung
    des Schmierstoffdruckes.
    IU-ETH-Zürich 1982

[9] Hora, P.; Reissner, J.: Werkstoffmechanisches Versagens-
    kriterium für duktile Blechwerkstoffe.
    Interner Forschungsbericht 1982, IU-ETH-Zürich

Simulation with Model Materials in the Nordic Countries

T. Wanheim, Laboratory for Mechanical Processes of Materials,
AMT-IPU, Technical University of Denmark, Lyngby/ Denmark

Summary

A series of projects on simulation metal-working processes
with model materials in a Nordic "action concertée" has been
carried through. The subjects dealt with are: Materials
properties, simulation of friction, plane simulation of
threedimensional problems, combined model material and computer simulation, applications in cold forging, drop forging,
rolling and extrusion.

A number of problems have been solved and a number of institutes and industrial plants in Denmark, Sweden, Norway and
Finland are now using the technique as a natural and valuable
help in developing metal-working processes, fixing proces parameters and designing tools.

Introduction

The use of waxes as model materials has proven its values in
numerous cases, and the reports on it go back several decades
/1/ - /6/.

In spite of this, the technique has been regarded with a
certain degree of scepticism as well in practical as in
academical circles.

Several years of working with the technique in research, con-
tract work and teaching has convinced the author that this
is a rather undeserved status for the idea /7/,/8/,/9/ and as
several other institutes in the Fenno-Scandinavian area
through the 70'ies showed growing interest in the field, it
was decided in -78 to coordinate this work and to make some
industrial application on a larger scale by carrying through
a set of projects in an "action concertée".

Pilot project

The first phase in the work was a survey project where the industrial needs for problem solving among the relevant processes were examined and the possible use of a physical modelling technique was assessed.

This was done by establishing a matrix where the relevant processes occupied one axis of the matrix and the aims of analysis that could be relevant to each proces on the other axis.

The list of processes is fairly obvious, see Fig. 1 and for the list of possible aims of analysis four groups were set up, problems relating to: the specimen, the interface between specimen and tool, the tool and the machine.

The idea was that the matrix should be used in three different ways to give answers to three fundamental types of questions:

- Which are the industrial problems within the processes mentioned and how critical are they?

- What is the "state of the art" of the wax modelling technique in the institutes working in the field in solving problems, i.e. which problems can be solved with the existing experimental know-how?

- Which future possibilities for problem solving can one foresee in the use of the wax simulation technique?

For each combination of process and aim of analysis in the matrices these questions were answered by assessing points (0-1-2-3) for graveness of need, level of experimental know-how and estimated possibility for application.

Aim of analysis			rolling						cold forging				hot forging				extrusion		drawing	
			hot rolling of slabs	hot rolling of profile	cold rolling of strip	wire rolling	tube rolling	transverse rolling	cold forging	warm forging	heading	ironing	hammer forging	drop forging	press forging	rotary forging	rod extrusion	tube extrusion	wire drawing	tube drawing
Specimen	filling of tool																			
	fibre structure																			
	strain distribution																			
	strain rate distribution																			
	hardness distribution																			
	temperature distribution																			
	obtainable deformation																			
	residual ductility																			
	residual stresses																			
	porosity																			
	spring back																			
Inter-face	friction																			
	normal stress																			
	sliding length																			
Tool	strain distribution																			
	stress distribution																			
	elastic deformation																			
	yield/fracture load																			
	temperature distribution																			
	thermal stresses																			
Machine	magnitude and direction of forces																			
	load displacement curve																			
	mechanical deformation																			
	thermal deformation																			

Fig. 1.   Matrix for pilot project.

The matrix for industrial need was in each of the four
countries worked out by institute personnel interviewing
leading industrial people in about 60 leading plants in
Denmark, Sweeden, Norway and Finland. The two other matrices
were answered by senior staff members in the institutes
working with the technique.

The SIMON project

The information collected in the pilot project was very
large, and the 12 projects chosen to be realized were ar-
ranged in three groups:

- applications

- modelling methods

- fundamental problems

The project was named SIMON, which is an abbreviation for
Simulation with Model Materials in the Nordic Countries,
5 different research laboratories have been involved with
about 10 researchers, some of whom have been doing their
thesis work.

Around 15 different industrial firms have participated, all
giving some kind of resources. The project has been super-
vised by a committee of 6 people, 4 leading proces development industrial engineers from Denmark, Finland, Norway
and Sweden, the head of the Copenhagen office for the
Nordic Co-operative Organisation for Applied Research
as administrative head and with a university representative
as chairman.

Of the project costs 50% were financed by The Nordic Countries'
Research Committee and 50% by national means, the latter
consisting of quite a volume of industrial money, the re-
minder coming from university and various research and

technology councils.

The publication of the project will be at three levels:

- Project reports (in Danish, Norwegian, Swedish and Finnish),
  mainly aiming at research laboratories and industrial
  specialists

- A compressed publication (approx. 50-70 pages) aiming at
  non-specialists and industrial people, stating the most
  important results of the project (each project described
  in one of the languages Danish, Swedish, Norwegian; Finnish translation of the text is provided)

- A series of papers in English - hopefully in the Journal
  of Mechanical Working Technology. The papers will range
  from rather short and practical results to a few more
  academical ones.

  The titles of the projects are the following:

Applications

- Model material simulation of non-ferrous hot forging (IVF)

- Designing preforming dies for drop forging using the
  model material technique (VTT)

- Simulation of cold forging folding problems using model
  materials (AMT/IPU)

- Model technique used in predicting the pressure
  distribution over the tool surfaces in cold forging (AMT/
  IPU)

- Extrusion of aluminium tubes using bridge dies (SINTEF)

- Designing extrusion dies by using the model material technique (VTT)

- Front and back end shapes formed in slab ingot rolling (KTH)

- Closing of porosities in stocks of square cross section by hot rolling with flat rolls (KTH)

Modelling methods

- Analysis of metal working processes using combined physical and numerical simulation (AMT/IPU - SINTEF)

- Plane model material studies of hot forging (IVF)

Fundamental problems

- Development of simulative tests for lubrication in model material experiments (AMT/IPU)

- Mechanical and thermal properties of model materials (VTT)

The letters in brackets are the national abbreviations for the research laboratory in charge of the project.

Thus:

AMT/IPU   :   The Technical University of Denmark

KTH   :   Royal Institute of Technology, Stockholm

IVF   :   The Swedish Institute of Production Engineering
                Research

SINTEF   : The Foundation of Scientific and Industrial
             Research at the University of Trondheim, Norway

VTT   : The Technical Research Centre of Finland.

<u>Description of Projects</u>.

<u>Simulation of non-ferrous hot forging</u>

(B. Anulf, IVF)

Hot forging is an efficient method for producing components
of brass with good tolerances and good mechanical properties.
In order to keep the costs sufficient low, the forging normally has to be carried out in one blow, and so the flow will
be very large and the material will move with high velocity
in many different directions. The risk for faults at corners
in the tools is therefore rather great, and the designer
often has to make several trials before an acceptable tool
geometry is found. This gives serious limitations in batch
sizes for this production method.

If model material experiments are to be carried out with any
degree of reliability, the material formability and friction
conditions must be under control. These parameters were
studied in this project together with the influence of the
tool velocity, air-inclusions in the tool and sensitivity
for changes in material composition, amount of lubricant
and temperature.

Tool velocity, temperature and material composition were
rather unimportant for the flow pattern of the specimen,
but the amount of lubricant has some influence.

For the cases studied graphite would be the natural lubricant
and the best agreement was obtained using talcum on the model.

As the yield stress of the model material is several hundred
times lower than the yield stress if the metal, airpockets
will have much greater influence in the model than in reality.

Holes for the escape of air must thus be bored in deep parts
of the tool or in corners. Therefore it is not acceptable to
use the production tool for model experiment, however, double
casting will easily produce a model tool.

In the study several cases of folding defects in brass
products were analysed and solved with model material
experiments.

Design of preforming dies for steel forging.

(K. Myrberg, S. Kivivouri, B. Ahlskog, VTT)

Of the total cost of a drop forged component 50% comes from
material used and 15% from material lost. By preforming the
component before the forging, the material loss can be reduced
and as the preforming in addition gives much better tool life
and possibilities of elimination forging defects, it is very
attractive.

In this study four different types of preforming were treated,
namely free forging, where the workpiece is moved and turned
and receives several blows, longitudinal rolling, where a
pair of rolls reduce the area of the workpiece in certain
places thereby elongating it and giving it the wanted geometry,
transverse rolling, where two plane tools move tangentially
to each other and finally open tool forging, where special
tool geometries with rotational and transverse movement
were investigated.

In all cases it was shown that use of the model material
technique yielded considerable savings in material, tool
life and eliminations of defects.

Simulation of cold forging folding problems

(J. Harnæs, AMT/IPU)

One of the serious limitations of cold forging is the folding
problems that occur e.g. when compressive stresses act in
slender parts of the specimen. To investigate whether the
model material technique is realiable for this type of problems a comparison has been carried through between folding
phenomena in steel, aluminium and model material.

The geometries chosen for this comparison were identical with
the geometries chosen by Dieterle. /10/.
In this way the experimental work was reduced and a reference
to an already published work was obtained.

If this comparison showed acceptable results, it would then
be a strong indication that the folding behaviour of modelmaterial specimens of more complicated geometries would be
a true indication of what could be expected in the real proces.

The proces limitations examined by Dieterle - the same as
were used in this investigation - were: fold limits for ring
forging, end-flange forging and tube-flange forging.
Results for model material specimens were found and compared
with Dieterle's results.

In all cases comparison showed that the modelmaterial folding
limits existed and were in good agreement with the folding
limits found for metal. The modelmaterial limits were in all
cases on "the safe side" of the metal limits.

If a cold forging proces in the design phase thus has been
investigated with the aid of a model-material experiment,
and no folds have been found, there are strong reasons to
believe, that the final proces in aluminium or steel will
run without folding problems.

Tool pressure distribution and tool stresses in cold forging

(V. Maegaard, AMT/IPU)

Since cold forging tools generally are subjected to very
high stresses the information about the pressure distribution over the tool-workpiece interface is important for
optimum tool design and choice of tool material.

In the present study two different but complementary
experimental techniques were used, namely as well determination of the pressure distribution over the specimen/tool
interface with model material and pressure transducers, as
determination of the stresses in a cold forging tool with
the aid of photoelastic models.

Measurement of the interface pressure in the tool was made
by membrane transducers of the strain-gauge type mounted
in transparent acrylate tools for rod an can cold forging.
Results showed that the pressure distribution is significantly
different from the often used hydrostatic approximation.

Investigation of the tool stresses was carried through with
the aid of photo-elastic tools, made by shrinking an inner
and an outer slice of the tool at a realistic temperature
difference. The stress distributions in tools with circular,
and square inner geometries were then determined in a number
of cases.

By loading the tool cavity with a rubber specimen forced to
expand by axial compression, working stresses in the tool
were modelled. With this technique rather complicated tool
geometries can be analysed using a realistic distribution
for the inner pressure in the tool.

Extrusion of aluminium tubes using bridge dies.

(H. Valberg, A. K. Hansen, SINTEF)

Aluminium tubes are often produced by extrusion with a bridge
die, i.e. a die, where the material has to flow on either side
of a number of "bridges" holding the mandrel and accordingly
weld together after the passage.

The flow pattern in this proces is complicated, and the
surface properties, mechanical properties of the tube and
the strength of the weld are dependent on the proces parameters.

The study of this type of proces is extremely difficult in
steel tools, and the model material technique seems to be
very promising for this application.

In the beginning of the present study focus was aimed at the
welding mechanisms in the welding chamber of the die. The
results showed that the welding problem was less serious
than expected, but that the surface conditions of the tube
were a major cause for unsatisfactory mechanical properties,
the test method being expansion of the tube by a conical
mandrel. The work was therefore directed towards the flow
in the die, as the flow and temperature conditions are of
major importance for surface quality.

Investigation of the dead-zone build-up showed that the "dead" material slowly was dragged out in the profile, giving rise to poor surface quality. Several bridge-geometries were investigated in metal and modelmaterial, and the quality of the weld behind the bridge tested.

Results showed that some dead-zone build-up is beneficial for the welding, but certain geometries can give problems in surface quality and thereby cause unsatisfactory mechanical properties of the tubes. The results were confirmed with experiments in aluminium.

Extrusion of hollow copperprofiles using mandrels.

(H. Pihlainen, S. Kivivouri, H. Kleemola, VTT)

Copper coils for electric generators are produced by extruding a profile which is cold drawn to final dimensions.

In extruding profiles with one or more holes for cooling water, mandrels have to be used, and if the pressure around the mandrel is not balanced, the mandrels will tend to bend in the proces, causing wall dimensions of the extruded products that lie outside the tolerances.

Two methods for experimental determination of the balanced position of the mandrels were studied.

In the first method a model material specimen with same geometry as the extruded profile is axially compressed between two plane tools with sticking friction.

In this compression the outer boundaries of the specimen will expand and assume an approximately circular contour when axial deformation is large enough. By drawing the successive contours of the deformation stages, a series of curves are obtained that describe a "natural" deformation from profile geometry to a circular contour. Assuming that these stages

also are valid for the reverse deformation gives then the
contours of the extrusion die at certain intervals.

The second method is based upon a hypothesis that the balanced
position for a mandrel or mandrels in the extrusion is the one
where the extrusion ratios on each side of the mandrel are
equal.

The planes in the material defining the ᵥ extrusion ratios
must be found experimentally.

This was done by dividing an extrusion billet in two at right
angles to the axis, paint a grid on one of the obtained surfaces, reassemble the billet and extrude it to some half-
way position.

After deformation the billet is opened and from the grid
deformation the flow-dividing planes can be found.

Both methods were succesfully applied in industrial extrusion
and are now part of the design-routines in the factory.

Front and back end shapes formed in slab ingot rolling.

(U. Ståhlberg, KTH)

The shape of overlaps formed at the front and back end in slab
ingot rolling is important for the materials economy. Thus
the material losses in cleaning front and back end amount to
10% by weight.

The volume of material that has to be removed is very dependent
on the geometric conditions in the rolling, and planning of
the scedule is accordingly important for the material utiliza-
tion.

In this project model materials were used to give the ideas

to and to refine a theoretical model for deformation in the
front and back end of the slab.

A theoretical model based on the upper bound method was derived
and refined with the aid of model material experiments.

Experimental verification was obtained with strain-hardened
aluminium slabs and very good agreement was obtained. Hot
steel was not used, in that the temperature gradients in the
ends would give deformation geometries that were not in agreement with the assumptions for the theoretical solution.

Results showed good agreement between theoretical model, model
material and aluminium. Results showed that the front-end fault
always was larger than the back-end fault and that large roll-
diameters and large reductions should be aimed at to minimize
material loss.

Closing of porosities in stocks of square cross section by hot rolling with flat rolls.

(A. Wallerö, KTH)

Steel in as-cast condition often contains central porosities.
This is mainly due to material shrinkage during solidification.
The central unsoundness has to be eliminated in the subseqent
process of hot working to ensure high yield and full strength
of the product.

Closing of porosities by hot working has been of great interest for a long time, especially considering tool and high-
alloy steels. In recent years this interest in porosity closure
has increased because of the replacement of conventional casting by continous casting. The crosssection of a continously
cast strand is usually small in comparison with that of an
ingot and consequently at the present time porosity closure
is often required to take place for even small reductions.

In this work the closure of a central porosity was analysed
for the process of hot rolling between flat and parallel
rolls. The workpiece studied has a square cross-section. A
three-dimensional upper bound solution was proposed in order
to analyse the closing of porosities. The theoretical analysis
was completed by experiments carried out with steel and model
material.

The experiments were performed with specimens furnished
with an artificial cylindrical hole in the centre. The gradual
change in shape of the hole was evaluated for rolling according to various rolling schedules.

The main conclusions from this work was that a rolling schedule
considering powerful closure of a central porosity shall be
built on heavy reductions by means of large rolls. Another
important conclusion was that rolling conditions resulting
in large spread are favourable. Small cylindrical holes are
more easily closed than large ones. However, also small holes
are difficult to eliminate by rolling. Carbon steel and stainless steel show, according to this work, no significant difference in the closure of a porosity. Comparing experiments it
becomes clear that the transformation of results obtained for
model material specimens to those of steel is very difficult
since the former spring back after deformation. Results obtained by model material experiments can in this context only be
used qualitatively.

Combined physical and numerical simulation

(M. P. Schreiber, AMT/IPU, T. Svinning, J. O. Løland, SINTEF)

Numerical simulation and physical simulation are complementary methods, each has its advantages and drawbacks.

The advantage of the computer-simulation is the ability to
calculate strains, stresses, temperature gradients etc., if

only the inputs regarding filling of the tool, direction and
magnitude of frictional forces and interface pressures between
tool and workpiece are stated. Up to now the computers have
only been able to generate these data in rather simple cases.

The advantage of the model material technique is that it physically illustrates how the deformation will proceed, and that
it is relatively easy to get the necessary starting data to
an advanced proces analysis from it. It is therefore close at
hand to try to combine these to methods to a powerful tool.

Three cases have been looked into in this study.

For a cold forging proces the model eksperiment has given
information on the flow in the tool, dead zone formations
etc., and the computer has from this basis calculated stresses in the tool and deformation forces.

For an extrusion proces the temperature rise has been calculated based on the flow pattern from the model experiment.
The temperature calculation is dependent on deformation,
friction, cooling against the die and the yield stress, which
it self is temperature dependant.

For hot rolling processes the forces and torques measured in
model experiments have been used to "calibrate" theoretical
calculations, thereby improving the accuracy considerably.

Plane studies of hot forging with axial symmetry

(B. Anulf, IVF)

Model studies of components with axial symmetry can be considerably simplified if a way could be found that would permit
the study of a plane strain analogy in stead of the true axial
case. The plane experiment is cheaper and quicker to carry out
and presents several advantages compared with the axisymmetrical case.

The aim of this project has been to investigate whether a
simple equation is capable of describing the deformation
relations between plane and axisymmetric deformation. The
equations in question is

$$x_p = \left(\frac{R_r}{R_o}\right)^2$$

and

$$y_p = y_r$$

where x_p is the horisontal distance of the point in question
from the axis of symmetry, R_r is the distance from center in
the real case, y_p and y_r are the vertical coordinates in the
plane and the real (axi-symmetric) case; R_O is a constant
influencing the outer profile of the plane model and its
radii and angles.

If a good picture of the material flow at radius R_m is wanted
experiments should be carried out with $R_O = 2R_m$, but results
from several cases showed that folding problems were better
simulated using $R_O = 2.5\ R_f$, where R_f is the radial distance
from centerline to fold in the real case.

If the fold in the real case is close to the centerline,
it is difficult to follow the above mentioned recommendation
as this will result in very long model tools.

Axisymmetric experiments are to be preferred in such cases.

Plane simulation of axisymmetric flow should always be followed by axisymmetric verification of the final geometry before
production tools are made. The plane simulation gives a means
for quick and easy tests on alternative geometries in the initial
phases, but should normally not be used alone.

Development of simulative tests for lubrication in model
material experiments

(N. Frederiksen, AMT/IPU)

This project includes a development of methods for simulation of the frictional conditions found in metal working
processes.

The simulation is carried out with carefully selected geometries from the four types: cold forging, drop forging,
extrusion and rolling.

The criterion used in selecting the geometries has been
their ability to show changes of frictional conditions by
changes in flow-pattern.

These geometries are then used as those measuring methods
that are going to ensure that the frictional conditions in
the simulation are the same as the frictional conditions in
the real process.

For cold forging a combined forward and backward can extrusion has been selected as an indication of the frictional condition.

For drop forging specimens with spike and with and without
flash has been investigated and geometry of test specimen has
been fixed.

For extrusion the conditions of the dead zones has been the
criterion for comparison between model and real proces.

Finally, looking at rolling, the phenomenon selected to
indicate if friction conditions are in agreement is the
forward slip.

Mechanical properties of model materials

(S. Finér, S. Kivivouri, H. Kleemola, VTT)

The stress-strain curve of the model material is important for the simulation. The constitutive equations of metals are of the form

$$\bar{\sigma} = f\left(\bar{\varepsilon}, \dot{\bar{\varepsilon}}, T\right)$$

where

$\bar{\sigma}$ is the yield stress
$\bar{\varepsilon}$ - - equivalent strain
$\dot{\bar{\varepsilon}}$ - - equivalent strain rate
T - - absolute temperature

The constitutive equations of the model material should be of the same form, even if full agreement of constants in the equation not necessarily is needed for all applications.

Stress strain curves for the material "FILIA" alloyed with varying amounts of kaoline, resin and microcrystalline wax were found by axial compression of cylindrical specimens.

The stress-strain curves were determined for various temperatures and strain rates.

The results were used to determine the constants in the selected constitutive equation

$$\bar{\sigma} = \sigma_0\left(\frac{\bar{\varepsilon}}{\bar{\varepsilon}_0}\right)^n \left(\frac{\dot{\bar{\varepsilon}}}{\dot{\bar{\varepsilon}}_0}\right)^m \left(1 + \beta \Delta T\right)$$

For pure FILIA the equation can be written

$$\bar{\sigma} = 0{,}69\,\bar{\varepsilon}^{\,0{,}12}\,\dot{\bar{\varepsilon}}^{\,0{,}18}\left(1 - 0{,}027\,\Delta T\right) \qquad 0 < \varepsilon \leqslant 0{,}3$$

and

$$\bar{\sigma} = 0{,}51\,\bar{\varepsilon}^{-0{,}12}\,\dot{\bar{\varepsilon}}^{\,0{,}18}\,(1 - 0{,}027\,\Delta T) \qquad 0{,}3 < \bar{\varepsilon} \leq 1$$

with ΔT in degrees centrigrade and $\bar{\sigma}$ in N/mm^2.

Constant for materials alloyed with different amounts of the ingredients mentioned have also been found.

References

/1/ Green, A. P.: The Use of Plasticine Models to Simulate the Plastic Flow of Metals. Philosophical Magazine, **42**, 327, 1951.

/2/ Bodsworth, C., Halling, I., Barton I. W.: The Use of Paraffine Wax as a Model Material to Simulate the Plastic Deformation of Metals. J. Iron Steel Inst., **185**, 1957.

/3/ Chang, K. T., Brittain, T. M.: An Investigation of Analog Materiales for the Study of Deformations in Metal Processing Simulations. ASME paper 67-WA/Prod-9, 1968.

/4/ Altan, T., Henning, H. J., Sabroff, A. M.: The Use of Model Materials in Predicting Forming Loads in Metalworking. J. Engineering Industry, **5**, 1970.

/5/ Sandin, A.: The Origin of the Exstrusion Defect in Copper and α/βBrass. Journ. Inst.Met., **90**, p. 439. 1961.

/6/ Brill, K.: Modellwerkstoffe für die Massivumformung von Metallen. Fakultät für Maschinenwesen der Technischen Hochschule Hannover, 1963.

/7/ T. Wanheim: Anwendung der Modelltechnik bei Massivumformgängen. Seminar "Neure Entwidklungen im Bereich der Massivumformen". Stuttgart. Industrie- Anzeiger 70 (1977).

/8/   T. Wanheim, M. P. Schreiber, J. Grønbæk, J. Danckert:
        Physical Modelling of Metal Forming Processes.
        Proces Modelling-Fundamentals and Applications ASM,
        1980, pp 145-165.

/9/   Model Technique in Metal Working Processes. Film, Labo-
        ratory for Mechnical Working of Materials, AMT, Techni-
        cal University of Denmark 1982.

/10/   Dieterle, K.: Faltenbildung als Verfahrensgrenze
        beim stauchen von Hohlkörpern. Berichte aus dem Insti-
        tut für Umformtechnik, nr. 30 Universität Stuttgart 1975.

Einfluß von Wechseln der Beanspruchungsrichtung auf die Fließspannung von Metallen

H. P. Stüwe, Erich-Schmid Institut für Festkörperphysik der Österr. Akademie der Wissenschaften, Leoben/Austria

Summary

Partial deformations in only one strain mode may be simply added to yield total strain. Reversal of the strain mode yields the Bauschinger effect. Repeated reversals lead to cyclic stress-strain curves and to a steady-state yield stress. Application to the distribution of hardness in swaged rods.

1 Addition von Teilverformungen im gleichen Sinn

Bild 1 zeigt schematisch eine typische Fließkurve. Sie beginnt mit der elastischen Geraden. Nach der Streckgrenze, die oft nicht deutlich markiert ist, verformt sich das Metall plastisch; die Fließkurve hat von hier an einen ungefähr parabolischen Verlauf.

Entlastet man die Probe im Punkt P, so geht die elastische Dehnung auf 0 zurück: es bleibt die plastische Dehnung. Belastet man erneut, so reicht die elastische Gerade bis zum Punkt P', der also die Streckgrenze des verformten Werkstoffes darstellt. Von da an setzt sich die Fließkurve fast so fort, als sei der Versuch nie unterbrochen worden. Auch die Punkte P und P' liegen gewöhnlich sehr nahe beieinander. In guter Näherung kann man also behaupten, die Fließkurve sei der geometrische Ort aller Streckgrenzen nach verschiedenen Verformungsgraden. Setzt sich die Verformung aus verschiedenen Teilschritten zusammen, so bleibt die Gleichung der Fließkurve unverändert, man kann schreiben:

$$\sigma = \sigma(\varphi) \qquad \varphi = \sum_i \Delta\varphi_i \qquad (1)$$

Dies gilt jedoch nur, wenn alle Schritte der Verformung im
gleichen Sinn erfolgen.

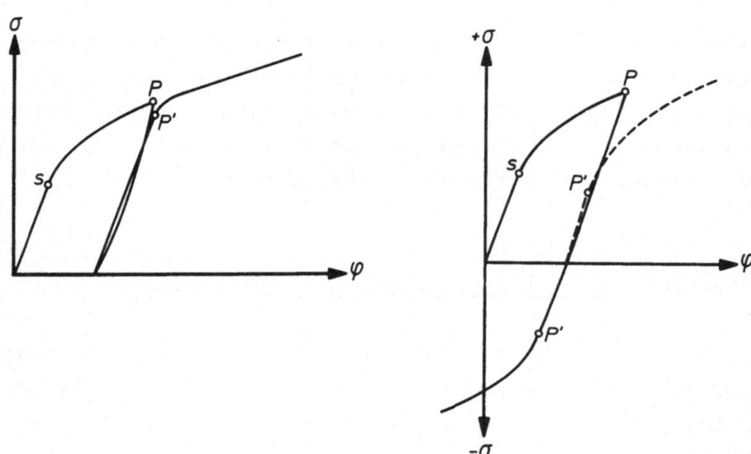

Bild 1: Fließkurve eines
Zugversuches mit Entla-
stung und erneuter Bela-
stung

Bild 2: Bauschinger -
Effekt

2   Fließspannung bei einmaliger Umkehr der Belastungsrichtung

Bild 2 zeigt den gleichen Versuch wie Bild 1, nur ist nach der
Entlastung die Belastungsrichtung umgekehrt worden. Dreht man
den unteren Teil der Kurve nach oben, so fallen die Punkte P und
P' nicht mehr zusammen und der weitere Kurvenverlauf kann nicht
mehr als Fortsetzung des anfänglichen Kurvenverlaufs aufgefaßt
werden. Diese Erscheinung ist längst bekannt unter dem Namen
"Bauschinger-Effekt".

Das bis zum Punkt P vorverformte Material hat also verschiedene
Streckgrenzen im Zug- und im Druckversuch. Diese Unsymmetrie
wird erklärt durch die Wirkung der Eigenspannungen, die bei der
Vorverformung erzeugt wurden. Bild 3 zeigt die Fließortkurve
eines Materials, das eine ähnliche Asymmetrie auch ohne Vorverformung zeigt: Die Streckgrenzen im Zug- und Stauchversuch sind
ganz verschieden. Hier rührt die Asymmetrie daher, daß die Verformung zum wesentlichen Teil von Zwillingsbildung getragen
wird. Bei einer einzelnen Versetzung ist es unwesentlich, ob
die an ihr angreifende Schubspannung durch einen Zug- oder einen
Stauchversuch erzeugt wird; lediglich die Bewegungsrichtung der
Versetzung kehrt sich um. Im Gegensatz dazu können Zwillinge in
einem bestimmten System nur in einer Richtung gebildet werden,
nicht aber in der Gegenrichtung.

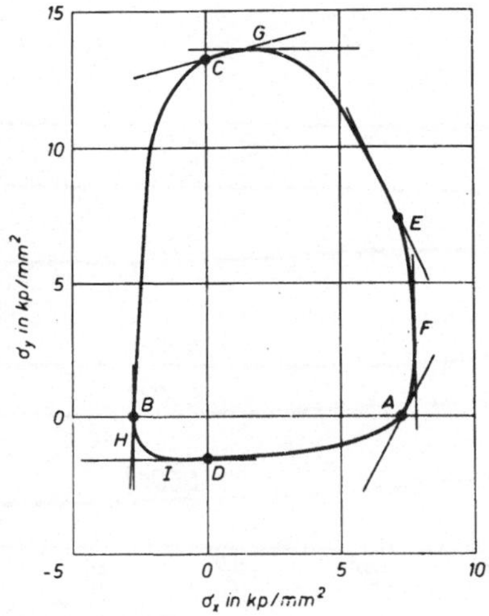

Bild 3: Kurve des Fließbe-
ginns für ein Magnesiumblech
mit Textur. Nach [1].

Diese beiden Beispiele zeigen, daß die Fließspannung nicht immer
- wie in der Plastomechanik häufig üblich - als Skalar behandelt
werden darf: die Fließspannung hängt vielmehr oft von der Art
der Verformung ab.

3   Zyklische Spannungsdehnungskurven

Wiederholt man Belastungszyklen wie im Bild 2 viele Male, so
kann man die Spitzenwerte der Punkte P durch eine einhüllende
Linie verbinden. Solche Kurven nennt man "Zyklische Spannungs-
dehnungskurven"; sie sehen so aus wie im Bild 4. Charakteristisch für solche Kurven ist ein ausgedehnter, ungefähr line-
arer Teil. Er kann ansteigen oder abfallen (Verfestigung und
Entfestigung); in sehr vielen Fällen verläuft er jedoch horizon-
tal, d.h. das Material erreicht einen stationären Zustand, in
dem sich die Fließspannung mit weiterer Verformung nicht mehr
ändert. Die Höhe des erreichten Plateaus ist in der Regel eine
Funktion der Amplitude der Verformung; bei höheren Verformungsgraden liegt das Plateau höher, bei niedrigeren Verformungsgraden liegt es tiefer.

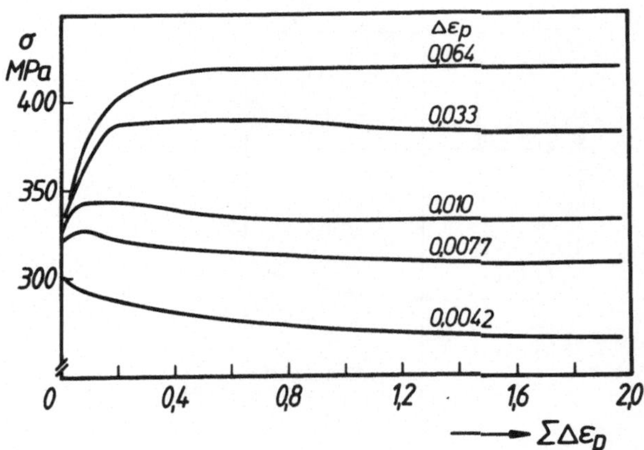

Bild 4: Wechselverformungskurve von kalt-
verfestigtem Cu bei 78K, Vorverformung 5%
bei 3OOK. Nach [2].

Bild 5 zeigt, daß diese stationäre Wechselfließspannung offenbar
einer Art Gleichgewichtszustand des Werkstoffes entspricht. Wird
nämlich die Verformungsamplitude oder auch die Temperatur des
Werkstoffs während des Versuchs erhöht oder erniedrigt, so ver-

ändert sich der Werkstoff und die neue Gleichgewichtsspannung
stellt sich ein.

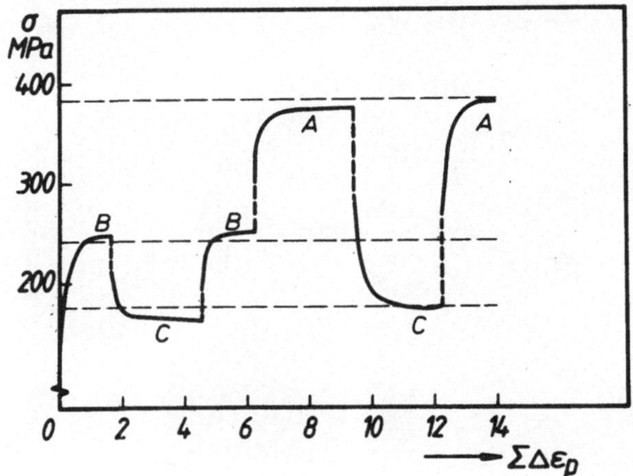

Bild 5: Einfluß des Wechsels der Verform-
ungsamplitude und / oder Temperatur auf
die Sättigungsspannung von Cu.
A: $\Delta\epsilon$ = 0,05, T = 78K; B: $\Delta\epsilon$ = 0,05,
T = 300K; C: $\Delta\epsilon$ = 0,005, T = 300K.
Nach [2].

4   Gefügebild

Die Bilder 6a und 6b zeigen zwei typische Gefügeformen, die sich
nach langer zyklischer Belastung ausbilden. Bei Bild 6a spricht
man von der sogenannten "Zellstruktur". Der Kristall besteht aus
Zellen niedriger Versetzungsdichte, die voneinander durch
Dickichte von zahlreichen Versetzungen getrennt sind. In Bild 6b
sieht man Subkörner, die ebenfalls nur wenige Versetzungen enthalten; sie sind voneinander getrennt durch klar ausgebildete
Subkorngrenzen.

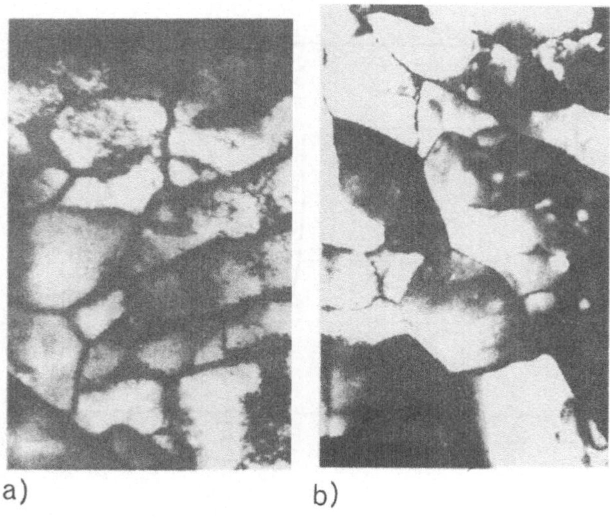

Bild 6: Versetzungsstruktur von wechsel-
verformtem Cu bei verschiedenen Spannungs-
amplituden σa. a) σa = 105 MPa.
b) σa = 130 MPa. Nach [3].

Der Beitrag der Verformungsversetzungen zur Fließspannung läßt
sich üblicherweise berechnen nach der Gleichung:

$$\sigma = AGb\sqrt{\rho} \qquad (2)$$

Darin bedeutet ρ die Versetzungsdichte. Bei einem Gefüge wie im
Bild 6a muß man mit einer mittleren Versetzungsdichte rechnen,
also mit dem Mittelwert der Versetzungszahl in einem Volumen,
das jedenfalls deutlich größer ist als eine Zelle. Bei einem
Subkorngefüge wie im Bild 6b ist es nicht möglich, sichtbare
Versetzungen zu zählen. Dennoch kann man aus der Zahl der Subkorngrenzen und ihrem mittleren Desorientierungswinkel eine Versetzungszahl abschätzen, die in Gl. (2) einzusetzen ist. Zu
ähnlichen Ergebnissen gelangt man, wenn man den Einfluß der Sub-

korngrenzen auf die Festigkeit nach einer "Hall-Petch-Beziehung" abschätzt, also:

$$\sigma = \sigma_0 + k_y\, d^{-\frac{1}{2}} \qquad (3)$$

Der mittlere Desorientierungswinkel geht dann in die Konstante k_y ein.

Es hat nun den Anschein, als ob die Menge der Versetzungen in den Zellwänden oder der Desorientierungswinkel in den Subkorngrenzen nicht sehr stark von der Vorgeschichte des stark verformten Materials abhängt. Die unterschiedlichen stationären Zustände, die im vorigen Abschnitt besprochen wurden, unterscheiden sich vielmehr vor allem durch die mittlere Größe der Zellen bzw. der Subkörner. Kleine Zellen bzw. Subkörner sorgen also für eine hohe Fließspannung, große Zellen und Subkörner entsprechen einer niedrigen stationären Fließspannung.

Bei weiterer Verformung verändern sich die Gefüge wie im Bild 6a und 6b nicht mehr wesentlich. Das bedeutet, daß genauso viele Versetzungen vernichtet wie erzeugt werden. Im Prinzip ist es zwar denkbar, daß bei Hin- und Rückverformungen jeweils die gleichen Versetzungen hin- und zurückbewegt werden, sodaß sich die Versetzungsdichte nicht zu ändern braucht. Diesen Fall gibt es auch tatsächlich bei sehr kleinen Verformungsamplituden (weit unterhalb von 1%). Bei höheren Verformungsgraden muß man aber annehmen, daß in jedem Verformungsschritt Versetzungsquellen betätigt werden und Versetzungen erzeugen. Um eine stationäre Zell- oder Subkornstruktur zu halten, muß deshalb die gleiche Anzahl von Versetzungen durch dynamische Erholung vernichtet werden. Die Zellgröße und damit der mittlere Laufweg der Versetzungen nimmt offenbar gerade den Wert an, bei dem dieses Fließgleichgewicht aufrecht erhalten werden kann.

Kompliziertere Verformungsfolgen

Über Wechselfließkurven nach Art von Bild 4 gibt es relativ
viele Untersuchungen, vor allem im Zusammenhang mit der Plasto-
ermüdung ("high-strain-fatigue"). Dagegen gibt es relativ
wenige Untersuchungen über kompliziertere Verformungsabläufe.
Ein besonders schönes Beispiel bieten die Untersuchungen von
Armstrong, Hockett und Sherby [4]. Sie haben würfelförmige
Proben wiederholt nacheinander in den drei Würfelachsen gestau-
cht und die entstehenden Gefüge untersucht. Die Ergebnisse
unterscheiden sich nicht sehr wesentlich von den zahlreichen
Versuchen mit einfacher Umkehr der Belastungsrichtung.

In diesem Beispiel scheint es sich allerdings um ein Material
ohne ausgeprägte Textur zu handeln; die Proben waren also zu
Beginn der Verformung quasi isotrop. In einem Material mit
einer stark ausgeprägten Textur - so wie etwa das Material im
Bild 3 - müßte eine solche Folge von Verformungszyklen zu sehr
unterschiedlichen Spitzenwerten der Fließspannung führen.

Anwendung auf das Rundhämmern

Bild 7: Rundgehämmerter Cu-Stab mit
Netzteilung. Nach [5].

Bild 7 zeigt einen Raster auf der Innenfläche einer geteilten
Stange, die durch Rundhämmern verjüngt wurde. Ganz ähnlich
sieht ein Raster auf einer geteilten gezogenen Stange aus. Es
ist üblich, aus der Verzerrung solcher Raster auf die Verformung zu schließen, die der Werkstoff erlitten hat [6]. Das ist
in diesem Fall jedoch unzulässig.

Bild 8 zeigt die Härteverteilung in einer gezogenen und in
einer rundgehämmerten Stange. Nimmt man die Härte - was zu-
lässig ist - als ungefähres Maß für die Fließspannung, so soll-
ten beide Stangen ungefähr das gleiche Profil zeigen: Eine
allgemeine Verfestigung als Folge der Querschnittsabnahme und
überlagert eine zusätzliche Verfestigung in den Randschichten
der Stange, hervorgerufen durch die Schiebung in der Nähe der
Werkzeugoberfläche. Diese Vermutung wird jedenfalls durch Bild 7
suggeriert. Die Härteverteilung in gezogenen Stangen sieht auch
tatsächlich so aus.

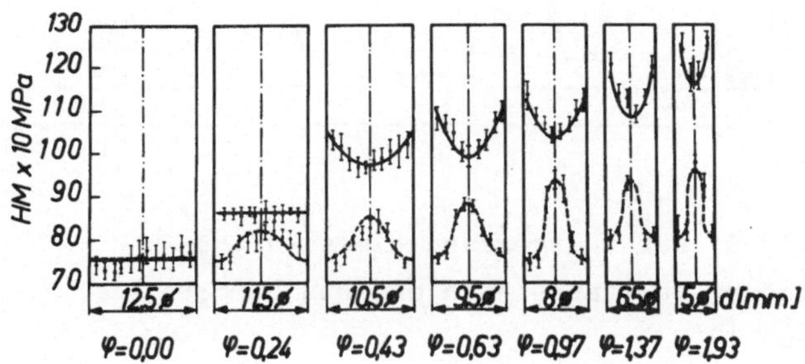

Bild 8: Härteverteilung in verformten
Cu-Stäben. (—— gezogene Stäbe, ----
rundgehämmerte Stäbe). Nach [5].

Die Härteverteilung der rundgeschmiedeten Stangen sieht jedoch ganz anders aus: Der Kern der Stange zeigt etwa jene Här-
tezunahme, die man nach der Querschnittsabnahme erwarten könnte,
der Rand aber ist weicher und zeigt kaum eine Verfestigung gegenüber dem Ausgangszustand.

Bild 9 zeigt diesen Effekt besonders deutlich: Eine gezogene
Stange, die anschließend durch Kalthämmern weiter verformt
wurde, erfährt durch die zusätzliche Verformung in den Randschichten keine Zunahme, sondern eine Abnahme der Festigkeit.

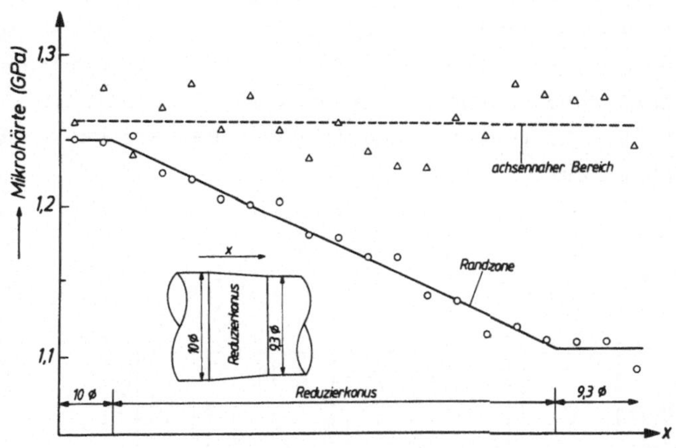

Bild 9: Härteverteilung eines kaltge-
zogenen Cu-Stabes nach dem Rundhämmern.
Nach [7].

Dieser Effekt kann nicht durch die einfache Addition von Verformungsgraden nach Gl. (1) erklärt werden, wohl aber durch die
Eigenschaften der Wechselfließkurven. Beim Rundschmieden wird
der Stab - insbesondere seine Randschichten - nicht einsinnig
verformt, sondern mit wechselnder Beanspruchungsrichtung. Dadurch stellt sich in diesen Zonen jenes Gleichgewichtsgefüge
ein, das der Wechselbeanspruchung entspricht. Durch das Rundschmieden wird also in den Randschichten eine dynamische Erholung erzwungen. Gefügeuntersuchungen bestätigen diese Vermutung.

Schrifttum

[1]   Hosford, W.F. in: Texturen in Forschung und Praxis, Hrsg.
      v. J. Grewen und G. Wassermann. Springer Verlag (1969),
      414.

[2]   Feltner, C.E.; Laird, C.: Cyclic Stress-Strain Response
      of F.C.C. Metals and Alloys - I.Phenomenological Experi-
      ments. Acta Metallurgica 15 (1967), 1621.

[3]   Lensing, R.; Mayr, P.; Macherauch, E.: Der Einfluß der
      Versuchsfrequenz auf das Wechselverfestigungsverhalten
      von reinen Kupfer- und   - Kupfer-Zinn-Vielkristallen.
      Zeitschrift für Metallkunde. 69 (1978) 6, 394.

[4]   Armstrong, P.E.; Hockett, J.E.; Sherby, O.D.: Large Strain
      Multidirectional Deformation of 1100 Aluminium at 300K.
      Journal of the Mechanics and Physics of Solids. 30 (1982),
      37.

[5]   Grabianowski, A.; Danda, A.; Ortner, B.; Stüwe, H.P.:
      Different Work Hardening in Swaged and Drawn Copper.
      Mechanics Research Communications. 7 (1980), 125.

[6]   Binder, H.: Untersuchungen über das Verjüngen von zylin-
      drischen Vollkörpern. Berichte aus dem Institut für Um-
      formtechnik, Universität Stuttgart, Nr. 58, Berlin ...:
      Springer 1980.

[7]   Schrank, J.; Ortner, B.: nicht veröffentlicht.

Ich danke Herrn Dipl.Ing. J. Schrank für seine Hilfe bei der
Vorbereitung dieses Manuskriptes.

Die Bedeutung der Umformverfahren für die systematische Werkstoffauswahl

P. Huml, Metallernas Bearbetning

Kungliga Tekniska Högskolan Stockholm / Schweden

Summary

Selection of materials is an important part of all product design and development. In this contribution the needs for adequate and reliable data on the properties of materials being formed - both material properties of the final product as well as the behaviour during plastic processing - are discussed. The development of a data bank cointaining such data suited to design and material selection procedures and coupled to CAD-systems is proposed.

1   Einleitung

Die steigenden Anforderungen an die Produkte, an die Wirtschaftlichkeit sowohl des Konstruierens als auch der Fertigung und die ständig zunehmende Anzahl verfügbarer Werkstoffe haben die Werkstoffoptimierung zu einem der wichtigsten Bestandteile der konstruktiven Planung und der stofflichen Verwirklichung eines Erzeugnisses gemacht. Im Laufe des letzten Jahrzehntes wurden mehrere Systeme und Hilfsmittel zur Effektivisierung und Erleichterung des Konstruierens und der Bauteiloptimierung entwickelt, wie z. B. Konstruktionskataloge (1) und algorithmische Werkstoffauswahlverfahren (2). Es wurden Datenzentren geschaffen, wo z. B. mechanische Eigenschaften von Metallen und Legierungen, Daten über Bearbeitung von metallischen Werkstoffen u. ä. gesammelt, kritisch bewertet und Datensammlungen erstellt werden, wie z. B. die Schnittwerte -Datenbanken (3). Die Anwendung von CAD/CAM hat sich durchgesetzt (4) und zu einer stürmischen Entwicklung, hauptsächlich in den USA, geführt (5).
Diese Entwicklung hat den Bedarf an leicht zugänglicher und für algorithmische Auswahlverfahren angepasster Information über Eigenschaften und Werkstoffverhalten bei der Verarbeitung von verfügbaren Werkstoffen akzentuiert.

In dem vorliegenden Beitrag werden in diesem Zusammenhang Schlussfolgerun-
gen bezüglich der Umformtechnik gezogen und die Anreize zu neuen Forschungs-
und Entwicklungsaufgaben diskutiert.

2   Beanspruchungs- und fertigungsgerechte Werkstoffauswahl

Bei der Auswahl geeigneter Werkstoffe für die verschiedenen Bauteile werden
Anforderungen an die Anwendungseigenschaften (mechanische, thermische, chemische u.s.w.) und an die Fertigungseigenschaften (Verarbeitbarkeit, wie
z. B. Zerspanbarkeit, Umformeignung, Schweissbarkeit u. ä.) in eine sog.
Anforderungsmatrix zusammengestellt (6). Diese wird dann mit Eigenschaften
verfügbarer Werkstoffe verglichen und ein solcher Werkstoff ausgewählt, der
sowohl den Anforderungen genügt wie auch andere Kriterien (der Wirtschaft-'
lichkeit, Verfügbarkeit in gewünschter Form u. d. gl.) erfüllt.

Die Anwendungseigenschaften ergeben sich aus der Analyse der betrieblichen
Funktion des betrachteten Bauteiles. Auch wenn es manchmal schwierig sein
kann, die mechanischen und thermischen Beanspruchungen quantitativ zuverlässig zu ermitteln, stehen heute zahlreiche Berechnungsmetoden und fertige
Berechnungsprogramme zur Verfügung, welche die Festigkeits- und Temperatur-
berechnungen wesentlich erleichtern und die Beanspruchungen quantitativ zu
erfassen erlauben. Schwierigkeiten bezüglich der Anwendungseigenschaften
treten oft erst dann auf, wenn die ermittelten Beanspruchungen, wie z. B.
Verschleiss- und Korrosionsbeanspruchungen u. ä. in messbare Eigenschaften
umgesetzt werden sollen. Die Folge der unscharf definierten Beanspruchungen
sind entsprechend ungenau den eigentlichen Anforderungen beanspruchungsge-
recht geforderte Werkstoffkennwerte. Ausserdem werden fast ausschliesslich
solche Werkstoffeigenschaften bei den in Frage kommenden Werkstoffen in Betracht gezogen, die in einem bestimmten Herstellungszustand, nicht selten
nach einer umfangreichen und komplizierten Wärmebehandlung in geometrisch
verhältnissmässig einfachen Versuchskörpern erzielt werden können.
Deswegen müssen bei der beanspruchungsgerechten Werkstoffauswahl den örtlichen Gesamtbeanspruchungen an den hochbelasteten Stellen eines Bauteiles

systematisch durch entsprechende örtlichen Werkstoffzustände spezifische
komplexe Werkstoffwiderstandsfähigkeiten entgegengesetzt werden (7).
Die Anforderungen an den gesuchten Werkstoff, die sich aus der Fertigung
des Bauteiles ergeben, werden bei der Werkstoffauswahl oft nur hinsichtlich
der Verarbeitungseigenschaften berücksichtigt (Giessbarkeit, Umformeignung,
Zerspanbarkeit u. ä.). Um die Werkstoffauswahl nun sowohl beanspruchungsge-
recht als auch fertigungsgerecht durchzuführen, muss man auch berücksichtigen, dass nahezu alle Fertigungsprozesse die mechanischen, physikalischen
und elektrochemischen Eigenschaften der Werkstoffe mehr oder weniger stark
beeinflussen. Der Einfluss der Fertigung auf Endzustände der Bauteile durch
umformende Prozesse wird in der Regel bei der Werkstoffauswahl qualitativ
berücksichtigt: Verbesserung der Festigkeits- und Verschleisseigenschaften
in kaltverformten Bauteilen und in thermomechanisch behandelten Stählen.
Doch auch hier werden solche Eigenschaften oder Endzustände berücksichtigt,
die in geometrisch einfachen Prüfkörpern ermittelt werden, ohne dass die
Verformungsgeschichte (einschliesslich des örtlichen Wechselns der Beanspru-
chungsrichtungen) im umgeformten Teil beachtet wird.

Der Prozess der Werkstoffauswahl stellt dann eine Anwendung des allgemeinen
Systems der Entscheidungsfindung. Die zugängliche Information über verfügbare Werkstoffe und ihre Verarbeitung ist dabei von ausschlaggebender Bedeutung. Die Daten über verfügbare Werkstoffe und ihre Eigenschaften werden
von Herstellern, Anwendern, Forschungsinstitutionen und verschiedenen Fach-
gesellschaften gesammelt. Bezeichnend für diese Information ist grosser Um-
fang und die daraus folgende Unübersichtlichkeit, Unvollständigkeit und
nicht zuletzt Zersplitterung. Diese Information wird dann von dem Konstrukteur gebraucht, um einen geeigneten Werkstoff auszuwählen. Auf den Konstrukteur strömt aber eine Flut von Informationen ein. Deswegen soll diese Information leicht zugänglich und leicht anwendbar sein. Die Gefahr wird dabei
immer grösser, dass im Bestreben die Informationen zu vereinfachen und über-
sichtlich zu gestalten, die Datensammlungen von Werkstoffeigenschaften,
Werkstoffverhalten während der Verarbeitung und Angaben über den Einfluss
der Verarbeitung auf die Endzustände der Werkstoffe solche Informationen

nicht mehr beinhalten, welche von strategischer Bedeutung für die Werkstoffauswahl sind. Ausserdem bedingen verschiedene Fertigungsverfahren spezielle
Fertigungseigenschaften, die ergänzend oder auch widersprüchlich zu den An-
wendungsanforderungen sind. Die grosse und ständig wachsende Zahl verfügbarer
Werkstoffe und ihrer verschiedener Zustände macht es praktisch unmöglich,
sich auf die menschliche Speicherkapazität ergänzt mit Nachsschlagswerken
zu verlassen. Deswegen werden Daten verfügbarer Werkstoffe in Datenbanken
gespeichert. Eine schwere und wichtige Aufgabe hierbei ist die systematische
Auswertung von Daten über Fertigungseigenschaften, über die Wechselwirkung
zwischen diesen und Werkstoffeigenschaften und über den Einfluss von Fertigungsverfahren auf die Endzustände verarbeiteter Werkstoffe. Diese Aufgabe
ist ausserordentlich schwierig und umfangsreich besonders in Bezug auf die
Umformverfahren wegen weitgehend fehlender nummerischer Klassifizierung komplexer Umformeigenschaften und wegen der Einwirkung der einzelnen Prozesspa-
rametern auf Endzustände der umgeformten Werkstoffe.

Die wertvollen Informationen, die heute auf diesem Gebiet vorliegen, sind
oft nur in Veröffentlichungen und Büchern zugänglich, welche sich dem Kreis
der in Umformung tätigen Fachleute zuwenden, siehe z. B. (9)! Hier bedarf es
darum einer Systematisierung und Speicherung von entsprechenden Daten über
die Umformung, ihren Einfluss auf die Endzustände umgeformter Werkstoffe
und Anforderungen an die Umformeignung einzelnerWerkstoffe oder Werkstoffgruppen. Um diese Aufgabe auf dem Gebiet der plastischen Umformung zu erfül-
len und die numerischen Daten unmittelbar nutzbar zu machen, ist es notwendig, dass die kritische Auswahl, Bewertung und Zusammenstellung der Daten
samt Erstellung von kritisch bewerteten Datensammlungen dort erfolgt, wo
eine unmittelbare Beziehung zu diesem Fachgebiet besteht: nahe der Forschung
und Entwicklung auf dem Gebiet der Umformtechnik.

3   Der Einfluss der Umformverfahren auf die Werkstoffeigenschaften

In vielen Fällen, besonders in niedrig beanspruchten Konstruktionen und bei
Werkstoffen, deren Auswahl ausschliesslich bezüglich der Wirtschaftlichkeit

geschieht, kann der Einfluss der Umformverfahren auf die Funktionseigenschaften des Bauteiles nur qualitativ in Betracht gezogen oder gar vernachlässigt werden. Dagegen i Bauteilen in hochbeanspruchten Konstruktionen, die in Werkstoffen mit hoher Festigkeit ausgeführt werden, müssen doch Einflussnahmen der Fertigungsverfahren, auch der Umformung, beachtet werden. Die Eigenschaften dieser Werkstoffe werden wesentlich durch die Umformverfahren verändert und manchmal auch geschädigt.

Für die Werkstoffauswahl ist es wichtig zu wissen, in welchem Ausmass die Eigenschaften beeinflusst werden können und wie derartiger Einfluss durch verschiedene Umformverfahren bei der Werkstoffauswahl zu berücksichtigen ist.

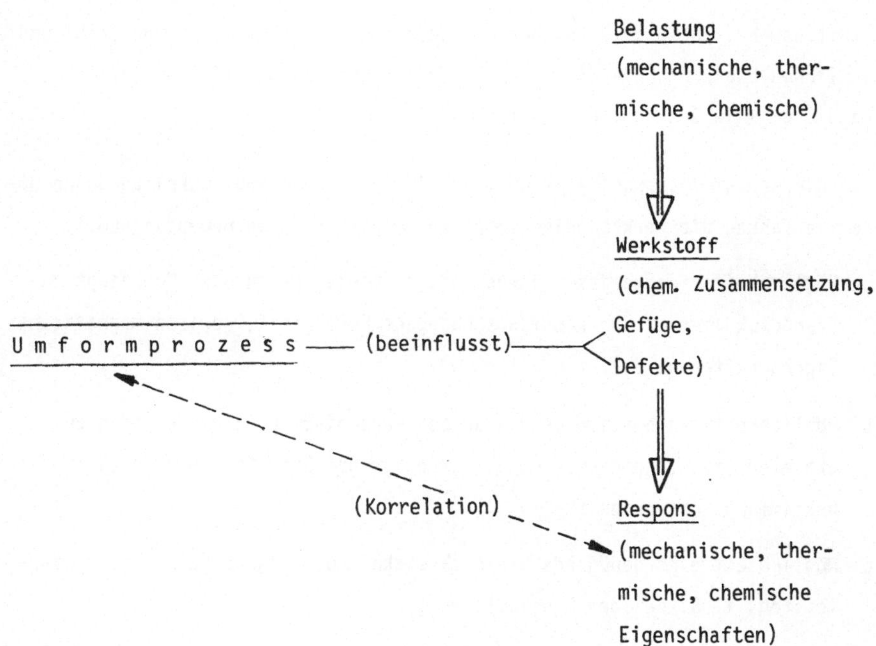

Fig. 1   Der Einfluss der Umformverfahren auf die Werkstoffeigenschaften

Wie in der Fig. 1 schematisch dargestellt ist, beeinflussen die Umformverfahren das Gefüge und die Defekte in den verarbeiteten Werkstoffen. Das, was
wir eine Eigenschaft nennen, ist das quantitative Mass von Respons, die
durch den Werkstoff (charakterisiert durch seine chemische Zusammensetzung)
und seinen Zustand (charakterisiert durch   Gefüge und Vorkommen von
Defekten) bestimmt wird. Dass man allgemein und besonders bei der Kaltumformung die Korrelation zwischen den Kenngrössen der Umformverfahren und End-
zuständen der verarbeiteten Werkstoffe als eine direkte Beziehung auffasst,
ist Folge der weitgehenden Aehnlichkeit zwischen verschiedenen kalt umgeformten Werkstoffen. Darüber hinaus ist es schwer und oft immer noch praktisch unmöglich, den Einfluss der Verarbeitung auf den Werkstoffzustand
zuverlässig quantitativ zu beschreiben, auch wenn viele neue Erkenntnisse
auf diesem Gebiet vorliegen (10, 11). Deswegen wird für die Zwecke der Werkstoffauswahl - besonders bei der Behandlung der Kaltumformung von Stahl und
ähnlichen Legierungen - die Korrelation anstatt des wirklichen Einflusses
nach Fig. 1 in Betracht gezogen.

Für die Werkstoffauswahl sind nun die vier folgenden Möglichkeiten durch Umformverfahren die Werkzeugeigenschaften zu beeinflussen bedeutungsvoll:

a. Möglichkeit eine gewisse Eigenschaft zu beeinflussen oder überhaupt zu-
   standezukommen (z. B. richtungsabhängige Festigkeit, gewisse magnetische
   Eigenschaften u. ä.)

b. Möglichkeit ein bestimmtes Niveau bei verschiedenen Eigenschaften zu
   erzielen (z. B. hohe Festigkeit: Streckgrenze über 1400 MPa in einem
   bestimmten rostfreien Stahl)

c. Möglichkeit eine genügende Gleichmässigkeit der Eigenschaften zu gewähr-
   leisten, z. B. bei der Blechumformung

d. Möglichkeit eine gewünschte Variation bestimmter Eigenschaften in einem
   Bauteil zu erzielen (z. B. über den Querschnitt des Bauteiles).

Diese Möglichkeiten, wenn sie gegeben sind, werden begrenzt teils durch die

Werkzeuge (wie z. B. durch die Festigkeit der Werkzeuge beim Kaltfliesspres-
sen, siehe (9, Bd. II)),teils durch Versagen des Verarbeiteten Werkstoffes
oder durch Beschädigung des Werkstoffes bei der Umformung mit Rücksicht auf
die Anforderungen an Funktionseigenschaften.

Die Möglichkeiten und Grenzen der Beeinflussung von mechanischen Werkstoffeigenschaften bei der Kaltumformung können oft den Fliesskurven entnommen
werden - mit dem Vorbehalt für die Uebertragbarkeit der Festigkeitswerte,
die an einfachen Prüfkörpern bei anderen Spannungszuständen ermittelt wurden
als bei solchen, die im jeweiligen Bauteil örtlich bei der Umformung herr-
schen, siehe auch (10)!
Weniger bekannt ist über den Einfluss verschiedener Verformungsverläufe bei
der Kaltumformung auf die Zähigkeit metallischer Werkstoffe. Hier bedarf es
einer kritischen Bewertung veröffentlichter Daten und einer systematischen
Zusammenstellung für Zwecke der Werkstoffauswahl.
Die Umformeignung verschiedener Werkstoffe zur Kaltumformung befindet sich
noch im Stadium des Suchens geeigneter Prüfverfahren, auch wenn beachtenswer-
te Ergebnisse   auf diesem Gebiet erzielt wurden, z. B. im Rahmen der
von STC Forming (C.I.R.P.) durchgeführten vergleichenden Untersuchungen der
Duktilität beim Kaltstauchen, oder am Institut für Umformtechnik der Universität Stuttgart, siehe (12)!
Sehr wenig ist bekannt über den Einfluss der Kaltumformung auf die anderen
Fertigungseigenschaften, hauptsächlich auf die Zerspanbarkeit. Die wenigen
Angaben über diese Einflussnahme sind - auch für ähnliche Zerspannungsver-
hältnisse-oft widersprüchlich, was auf streng werkstoffspezifische Einfluss-
mechanismen deutet. In diesem Fall scheint es zweckmässig zu sein, den Einfluss des Umformverlaufes auf das Gefüge des verarbeiteten Werkstoffes mit
in Betrachtung zu ziehen (Fig. 1).
Für die Auswahl von Werkstoffen für Tiefziehteile müssen neben Festigkeit und
den bekannten Werkstoffkennwerten für das Tiefziehen (R- und n-Werte) auch
solche Eigenschaften mitberücksichtigt werden wie z. B. Beulsteifigkeit und
Schallabstrahlung fertiger Tiefziehteile (13). Auch hier hat sich in den
letzten Jahren gezeigt, dass das Werkstoffverhalten bei der Blechumformung

zweckmässig mit Hilfe von metallphysikalischen Modellen beschrieben werden kann (14).

Was die Warmumformung betrifft, kan festgestellt werden, dass das Ziel für die Warmumformung, auch von Halbzeug, mit wenigen Ausnahmen bis vor wenigen Jahren nur die Herstellung von Erzeugnissen mit gewünschter Formen und Dimensionen gewesen war, während in den letzten etwa zehn Jahren immer öfter angestrebt wird die Endzustände der warmumgeformten Erzeugnisse den Anforderungen bereits während oder im Anschluss zu den Verfahren zu erhalten.
Der Einfluss der Warmumformung auf das Gefüge umgeformter Werkstoffe ist verhältnismässig gut beschrieben und heute ist es möglich, die Endzustände warmgeformter Werkstoffe vorauszusagen (15). Das wird sowohl beim Warmwalzen (15), beim Schmieden (13) und auch anderen Verfahren ausgenutzt, z. B. beim Vergüten aus der Umformwärme.
Auch über das Verhalten der Werkstoffe beim Warmumformen liegen Daten vor, die bereits teilweise in zugänglichen Datenbanken gesammelt wurden (16).
Die Bemühungen in diesem Zusammenhang sind heute gerichtet auf die Entwicklung teils von praktisch hantierbaren Fliesspannungsmodellen, siehe auch (11), teils von einfachen und zuverlässigen experimentellen Methoden zur Ermittlung der Warmumformbarkeit.
Die bei der Warmumformung auftretenden strukturellen und Gefügeänderungen müssen werkstoffspezifisch berücksichtigt werden und sind zum Unterschied von Kaltumformung so eng mit dem verarbeiteten Werkstoff verknüpft, dass sie die Werkstoffauswahl in dem Sinne bestimmen, dass die geforderten Eigenschaften im weiten Variationsbereich gezielt können erhalten werden.
Mangel an Daten in dieser Beziehung betrifft in erster Reihe Optimierung der Fertigungsverfahren, hauptsächlich hinsichtlich der Anforderungen an Gleichmässigkeit der Eigenschaften und weniger die Werkstoffauswahl als solche.

Halbwarmumformung gewinnt ständig an Bedeutung: das Umformvermögen ist bei den Temperaturen der Halbwarmumformung grösser, die Fliesspannung und demzufolge auch die Umformkräfte sind kleiner, während die halbwarm umgeformten

Werkstücke eine bessere Oberflächengüte und bessere Genauigkeit als warm-
umgeformte Bauteile aufweisen. Ueber die Eigenschaften halbwarm umgeformter
Werkstoffe ist doch verhältnismässig wenig bekannt. Einfluss dieser Umformverfahren auf die Endzustände der Erzeugnisse verdient wegen der technischen und wirtschaftlichen Bedeutung entsprechendes Interesse.

Metallpulverschmieden zeichnet sich nicht nur durch eine bessere Genauigkeit der Werkstücke aus, sondern auch durch eine bedeutend grössere Gleichmässigkeit der Eigenschaften und durch Möglichkeit solche Kombinationen
von Eigenschaften zu erzielen, die in herkömmlichen Verfahren nicht erhalten werden können (17), siehe auch (18)!
Für die Zwecke der Werkstoffauswahl bedarf es doch einer für die systematische Werkstoffauswahl geeigneten Zusammenstellung von möglichen Eigenschafts-
variationen, die in den pulvergeschmiedeten Werkstücken erzielt werden
können.

4   Auswahlmetoden: Kriterien und Voraussetzungen

In den letzten etwa fünfzehn Jahren wurde die Werkstoffauswahl systemati-
siert und es wurden entsprechende algorithmische Auswahlverfahren entwickelt
(2, 6, 7, 19).
Im eigentlichen Auswahlsvorgang werden die Anforderungen an die Eigenschaften der gesuchten Werkstoffe (zusammengestellt in der Anforderungsmatrix)
mit den Eigenschaften verfügbarer Werkstoffe verglichen. In einfachen Fällen,
wenn die Anforderungsmatrix widerspruchsfrei ist, wird die Werkstoffauswahl
zu einem einfachen Vergleich reduziert und der gesuchte Werkstoff kann dann
ohne Schwierigkeiten gefunden werden. So ist es doch nicht der Fall, wenn
die Anforderungsmatrix komplex wird, oder wenn die Anforderungsmatrix wider-
sprechende und/oder unscharf definierte Anforderungen enthält. In solchen
Fällen müssen dann die Anforderungen ranggeordnet und gewichtet werden.
Die Rangordnung erfolgt nach entweder ausschliesslich technischen, oder
technisch-wirtschaftlichen, rein wirtschaftlichen oder gegebenfalls nach
anderen (z. B. umweltschutzgesetzlichen) Kriterien. Die Auswahl nach dem
für den betrachteten Fall relevanten Kriterium geschieht dann meistens unter

wertanalytischen Gesichtspunkten, siehe z. B. (2).

Die Durchführung der algorithmischen Werkstoffauswahl - bis auf die erwähnten einfachen Fälle - ist ohne maschinell zugängliche Datenbanken praktisch fast unmöglich. In den Unternehmen, in welchen die systematische Werkstoffauswahl realisiert wurde und laufend ausgenutzt wird, hat sich erwie-
sen, dass die wahrscheinlich wichtigste Voraussetzung für die praktische
Bewährung dieser systematischen Verfahren die rechnerunterstützten Methoden
und hauptsächlich zuverlässige und umfangreiche Datenbanken sind (20).
Für diese Zwecke wurde 1978 in den USA von Metal Properties Council die Erstellung einer umfangreichen Datenbank gestartet (21), in welcher die mechanischen Eigenschften metallischer Werkstoffe gespeichert werden. Aehnli-
che Datenbanken (Mechanical Properties Data Center und Metals and Ceramics
Information Center) wurden von Battelle Memorial Institute in Columbus, Ohio,
erstellt. Werkstoffdaten zu Stahl und Eisen werden in einer Datenbank am
Betriebforschungsinstitut (VDEh) in Düsseldorf, Zerspannungsdaten am INFOS
in Aachen und bei MDC, Metcut R. A. in Cincinnati, Ohio, gespeichert (22).

Wie oben mehrmals betont wurde, ist es für die optimale Werkstoffauswahl
notwendig, eben den Einfluss der Umformverfahren auf Endzustände der umgeformten Werkstoffe mitzuberücksichtigen. Diese Voraussetzung kann nur dann
erfüllt werden, wenn entsprechende Daten, auch über die Möglichkeiten und
Begrenzungen solcher Einflussnahmen, dem Konstrukteur zugänglich gemacht
werden.

5 Anpassung an die Anforderungen in CAD-Systemen

Für die Werkstoffauswahl ist der Konstrukteur verantwortlich, auch wenn der
Werkstoffingenieur in diesem Prozess mitwirkt. CAD hat sich als ein effekti-
ves Hilfsmittel in der konstruktiven Planung, in Produktionsplanungs- und
Vorbereitungsprozessen (rechnerunterstützte Herstellung von Werkzeugen)
praktisch bewährt. Die Anreize für den Einsatz von CAD liegen heute vor-
rangig in der Aufgabe, den kreativen Konstruktionsprozess wieder auf seine
wesentliche Elemente zurückzuführen (23). Die ständig wachsende Komplexität
und die grosse Anzahl gesetzlicher und marktbedingter Aspekte macht es

notwendig sich der rechnerunterstützten Methoden im grösseren Masse zu bedie-
nen, wie sich z. B. in den USA gezeigt hat, als es nötig war in kürzester
Zeit kleinere Automobilmodelle auf den Markt zu bringen. Es kann mit Sicherheit erwartet werden, dass Europa, das - was CAD betrifft -hinter den USA
liegt, eine ähnliche stürmische Entwicklung der CAD-Technologie bevorsteht
(23).

Die rechnerunterstützten Hilfsmittel der systematischen Werkstoffauswahl werden mit grosser Wahrscheinlichkeit innerhalb kurzer Zeit ein Bestandteil
von CAD-Systemen. Darum scheint es angebracht, in der nahen Zukunft die
Entwicklung der Datenbanken für Zwecke der systematischen Werkstoffauswahl
den Anforderungen der CAD-Systeme anzupassen.

6 Schlussfolgerungen

Die Werkstoffauswahl kann nur dann beanspruchungs- und fertigungsgerecht
durchgeführt werden, wenn auch der Einfluss der Umformverfahren auf die Endzustände verarbeiteter Werkstoffe berücksichtigt wird.
Für die Zwecke der systematischen Werkstoffauswahl sind rechnerunterstützte
Methoden und entsprechende Datenbanken von ausschlaggebender Bedeutung.

In den heute vorhandenen Datenbanken gibt es Lücken in Informationen über
die Kennwerte der Umformung, über die Möglichkeiten und Begrenzungen der
Einflussnahme verschiedener Umformverfahren auf das Verhalten bearbeiteter
Werkstoffe während der Umformung und auf die Endzustände umgeformter Bauteile.

Die Forschung und Entwicklung auf dem Gebiet der Umformtechnik kann einen
wertvollen Beitrag zur Entwicklung der Werkstoffauswahl und der CAD-Systeme
durch Erstellung von entsprechenden Datensammlungen bzw. Datenbanken leisten.
Dabei kann von der von K. Lange (9) stammenden Systematisierung der Umformverfahren ausgegangen werden.

- 212 -

Schrifttum

(1)   Roth, K.: Konstruieren mit Konstruktionskatalogen
              Springer Verlag, Berlin, New York, Heidelberg, 1981

(2)   Fulmer Research Institute: Fulmer Materials Optimizer, Vol. I - IV
              Stoke Poges, Buckinghamshire

(3)   INFOS: Schnittwerte
              Informationszentrum für Schnittwerte
              Laboratorium für Werkzeugmaschinen, Technische Hochschule
              Aachen

(4)   Spur, G.; Krause F.-L.: Aufbau und Einordnung von CAD-Systemen
              VDI-Berichte 413(1981), 1-18

(5)   McElroy, J.: CAD/CAM
              Automotive Industries   (1981):7, 35-38

(6)   Grosch, J.: Systematik der Werkstoffauswahl
              ZwF 68(1973):1, 26-32

(7)   Illgner, K.-H.: Beanspruchungsgerechte Werkstoffauswahl als Optimie-
              rungsaufgabe
              VDI-Berichte 410(1981), 1-14

(8)   Steffens, H.-D.: Fertigungsgerechte Werkstoffauswahl
              VDI-Berichte 410(1981), 15-27

(9)   Lange, K.: Lehrbuch der Umformtechnik, Bd. 1-3
              Springer Verlag, Berlin, New York, Heidelberg, 1974

(10) Stüwe, H.-P.: Der Einfluss von Wechseln der Beanspruchungsrichtung
              auf die Fliesspannung von Metallen
              Symposium "Grundlagen der Umformtechnik"
              Institut für Umformtechnik der Universität Stuttgart,
              13. und 14. Oktober 1983

(11) Steck, E.: Entwicklung von Stoffgesetzen für die Hochtemperaur-
              Plastizität
              Symposium wie in (10)

(12) Pöhlandt, K.: Ueber die Wechselwirkung von Werkstoff und Umformung
              Symposium wie in (10)

(13)   Doege, E.: Stand und Tendenzen in der Umformtechnik
           VDI-Z 123(1981):10, 385-394

(14)   Bergström, Y.: Flow Stress Models for Sheet Metal Forming
           KTH Stockholm 1981

(15)   Meyer, L.: Neuzeitliche Umform- und Wärmebehandlungsverfahren zur
           Erzielung günstiger Werkstoffeigenschaften
           VDI-Berichte 428(1981), 35-42

(16)   Roberts, W.: Flow Stress Data for Hot Forming (Evaluation and Use)
           Swedisch Institute for Metals Research, Stockholm 1983

(17)   Huppmann, W.-J.: Powder Forging
           Int. Metals Review 23(1978), 209-239

(18)   Wilhelm, H.: Turbinenscheiben für Flugtriebwerke
           Symposium wie in (10)

(19)   Grosch, J.: Werkstoffgerechtes Konstruieren und Werkstoffauswahl
           VDI-Berichte 385(1980), 1-8

(20)   Graham, J.-A.: Materials Selection and Evaluation at Deere & Co.
           Conference on Material Selection, Lund (Sweden), Sept.1979

(21)   Graham, J.-A.: Mechanical Properties Data Base
           Metal Properties Council, Houston, Texas, Sept. 1979

(22)   Hörz, G.: Informations- und Dokumentationswesen auf dem Gebiet
           der Metalle, Teil I und II
           Metall 33(1979):3, 274-284 und 34(1980):1, 54-60

(23)   Lincke, W.: Zukünftige CAD-Anwendungen, Forderungen und Perspektiven
           VDI-Berichte   413(1981), 137-142

Über die Wechselwirkung von Werkstoff und Umformung und ihre Beschreibung durch Versuche

K. Pöhlandt, Institut für Umformtechnik, Universität Stuttgart

Summary

Experiments are described and discussed for determining material properties before metal forming as well as process parameters during the metal forming operation and workpiece properties after metal forming.

Before metal forming pure properties of matter have to be determined which in many cases is achieved by standardized testing methods. After the forming process, however, in addition to material properties, the effect of geometry and the requirements of the particular function of the finished workpiece must be taken into account. Therefore, a great variety of testing methods is to be applied.

Verzeichnis der wichtigsten Abkürzungen

A_g	Gleichmaßdehnung
C	Konstante in Gl. (1)
ε	Dehnung
F	Kraft
φ	Umformgrad
φ_g	Umformgrad bei Gleichmaßdehnung
k_f	Fließspannung
n	Verfestigungsexponent
R_m	Zugfestigkeit

1 Einleitung

Unter der "Wechselwirkung zwischen Werkstoff und Umformung" wird im folgenden der Gesamtkomplex der gegenseitigen Abhängigkeit und Beeinflussung eines Werkstoffes - damit ist hier stets der Werkstückwerkstoff gemeint - mit einem Umformverfahren verstanden. Es soll also der Begriff der Wechselwirkung in einem sehr weiten Sinne aufgefaßt werden und nicht auf den eigentlichen Umformvorgang beschränkt bleiben.

In diesem weiteren Sinne beginnt die Wechselwirkung schon in dem Augenblick, in dem ein spezieller Werkstoff für ein spezielles Umformverfahren ausgewählt wird, weil er dafür geeignet erscheint, und sie endet nicht mit dem Ende des Umformvorganges, sondern äußert sich noch im Nachhinein in den Eigenschaften des gefertigten Werkstückes.

Die Gliederung des Vortrages ergibt sich aus einer Betrachtung der Umformtechnik als System, vgl. Bild 1 /1, 2/.

Den theoretischen Hintergrund für diese Betrachtungsweise liefert die Systemtheorie /3, 4/, die in der Umformtechnik bisher hauptsächlich auf tribologische Probleme angewandt wurde /5/. Im Rahmen des Vortrages brauchen allerdings systemtheoretische Fragen nicht erörtert zu werden. Auf eine mathematisch strenge Definition des Systembegriffes, wie sie in /4/ vorgelegt ist, kann verzichtet werden. Es genügt hier die folgende Definition:

"Ein System ist ein Satz von Elementen, die durch Struktur und Funktion miteinander verbunden sind" /5/.

In den folgenden Abschnitten werden nacheinander die Punkte 1 bis 4 in Bild 1 betrachtet.

Um das im Prinzip sehr breit angelegte Thema auf beschränktem Raum bzw. in beschränkter Zeit zu bewältigen, sind einige Einschränkungen notwendig.

Eine erste Einschränküng ergibt sich daraus, daß im Wesentlichen nur Volumenphänomene betrachtet werden sollen, d.h. Vorgänge oder Eigenschaften, die sich auf das Werkstoff- oder Werkstückvolumen beziehen (im Gegensatz zu Oberflächenphänomenen). Beim Umformvorgang konzentriert sich gewissermaßen

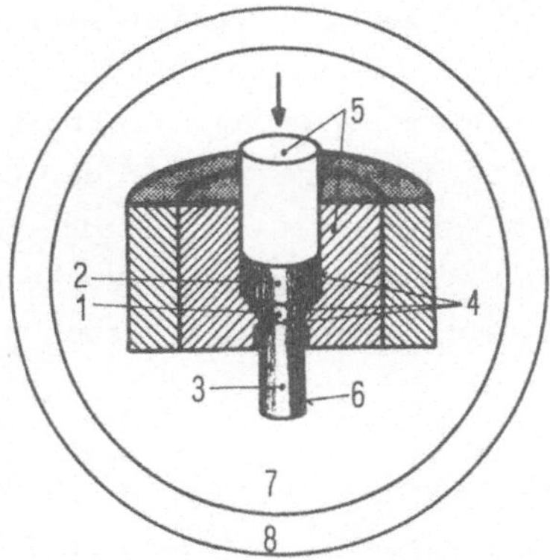

1. Umformzone
2. Stoffeigenschaften vor dem Umformen
3. Stoffeigenschaften nach dem Umformen
4. Wirkfuge zwischen Werkstück und Werkzeug
5. Umformwerkzeug
6. Oberflächenreaktionen zwischen Werkstück und umgebender Atmosphäre
7. Werkzeugmaschine
8. Betrieb

Bild 1. Das System der Umformtechnik am Beispiel des Fließpressens /1/

das Geschehen auf die Wirkfuge zwischen Werkstück und Werkzeug, so daß Oberflächenphänomene und tribologische Vorgänge eigentlich eine zentrale Rolle spielen. Da aber tribologische Probleme auch in anderen Tagungsbeiträgen behandelt werden, sollen sie hier ausgeklammert bleiben.

Eine weitere thematische Beschränkung ergibt sich daraus, daß hier lediglich eine Übersicht gegeben werden soll, bei der Einzelheiten außer Acht gelassen werden. Es wird kein Anspruch auf Vollständigkeit erhoben. Aus der Vielfalt der Umformverfahren werden exemplarisch einige wenige herausgegriffen. Die Aufzählung der zu prüfenden Werkstoff- bzw. Werkstückeigenschaften und der zugehörigen Versuche ist breiter angelegt, aber auch sie kann nicht erschöpfend sein.

Es wird zunächst der Werkstoff vor der Umformung betrachtet. Dabei kann vielfach vorausgesetzt werden, daß der Werkstoff im weichgeglühten Zustand vorliegt, so daß er den Umformgrad $\varphi = 0$ hat. Es werden die Versuche beschrieben, die eine Aussage über die Umformeignung des Werkstoffes gestatten, wobei auch werkstoffbedingte Verfahrensgrenzen zum Begriff der Umformeignung zu zählen sind. Daran schließt sich eine knappe Betrachtung des Umformvorganges an. Es folgt eine Betrachtung des gefertigten Werkstückes - dabei ist in erster Linie an die Stückgutfertigung gedacht. Hiermit wird von der Werkstoffprüfung auf die Werkstück- oder Bauteilprüfung übergegangen.

Es ist im folgenden stets zu bedenken - auch wenn dies nicht explizit hervorgehoben wird -, daß ein Versuch nicht losgelöst von der Theorie betrachtet werden darf: "erst die Theorie entscheidet darüber, was wir messen können" (Einstein). Nur mit Hilfe theoretischer Vorstellungen lassen sich Meßergebnisse deuten und verallgemeinerte Schlüsse ziehen.

2 Der Werkstoff vor der Umformung

2.1 Übersicht

Im folgenden kann in der Regel vorausgesetzt werden, daß das Rohmaterial bei der Blechumformung ein Blechwerkstoff ist, und bei der Massivumformung rotationssymmetrisches Stangenmaterial. Zur Charakterisierung eines Werkstoffes vor der Umformung (Punkt 1 in Bild 1) gehören alle Eigenschaften, die in Bild 2 zusammengestellt sind.

Diese Eigenschaften lassen sich unterteilen in solche, die für die Beurteilung des Umformverhaltens im engeren Sinne maßgeblich sind - in Bild 2 links oben -, und allgemeine Eigenschaften, die möglicherweise Rückschlüsse auf zu erwartende Eigenschaften des fertigen Werkstückes zulassen (rechts oben), oder die für die metallkundliche Deutung des Werkstoffverhaltens von Bedeutung sind (unten). Dabei gibt es Fälle, in denen eine Eigenschaft nicht eindeutig zugeordnet werden kann.

Es kann im folgenden grundsätzlich angenommen werden, daß der zu untersuchende Werkstoff zumindest in Näherung den an ihn zu stellenden Anforderungen genügt, d.h. es ist eine Vorauswahl getroffen worden, in der völlig ungeeignete Werkstoffe schon ausgeschlossen worden sind. Die zu beschreibenden Versuche dienen somit dazu, die - im Prinzip zu erwartende - Umformeignung eines Werkstoffes quantitativ zu bestimmen.

2.2 Versuche zur Beurteilung des Umformverhaltens

Zunächst werden die Eigenschaften der ersten Gruppe (Bild 2 links oben) betrachtet, die eine Aussage über das Umformverhalten des Werkstoffes ermöglichen. Hier sind von besonderem Interesse die Festigkeitskennwerte wie Streckgrenze und Zugfestigkeit, ferner die Fließkurve und die Zähigkeitskennwerte Bruchdehnung und Brucheinschnürung.

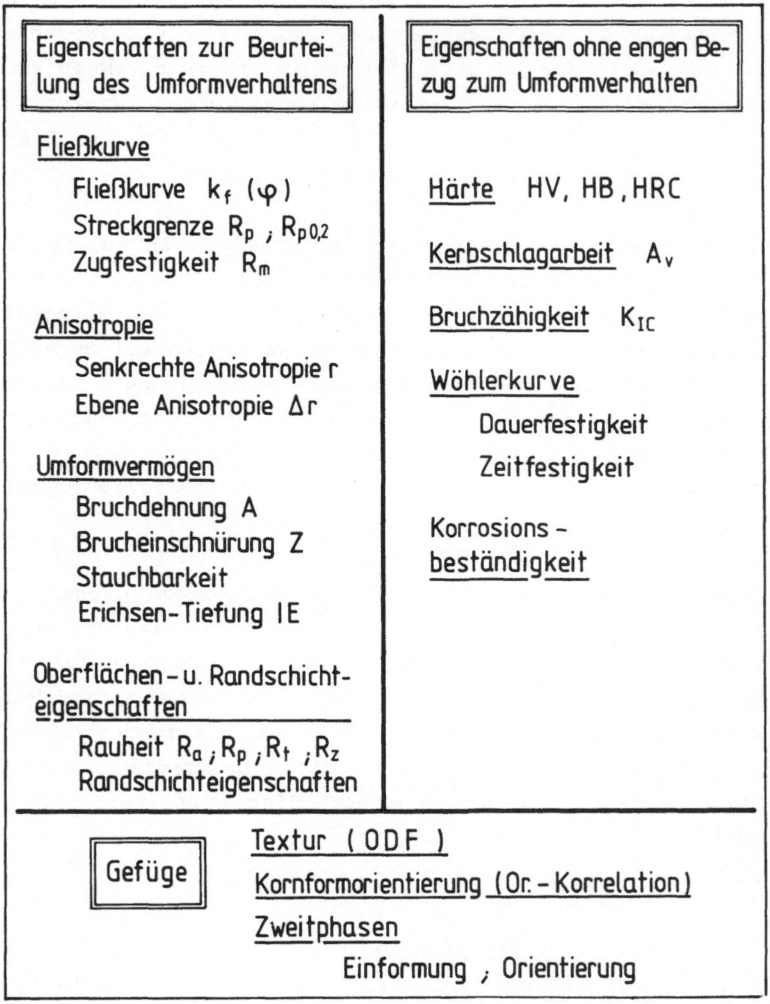

Eigenschaften zur Beurteilung des Umformverhaltens

Fließkurve

Fließkurve $k_f\,(\varphi)$
Streckgrenze R_p ; $R_{p\,0,2}$
Zugfestigkeit R_m

Anisotropie

Senkrechte Anisotropie r
Ebene Anisotropie Δr

Umformvermögen

Bruchdehnung A
Brucheinschnürung Z
Stauchbarkeit
Erichsen-Tiefung IE

Oberflächen – u. Randschichteigenschaften

Rauheit R_a ; R_p ; R_t ; R_z
Randschichteigenschaften

Eigenschaften ohne engen Bezug zum Umformverhalten

Härte   HV, HB , HRC

Kerbschlagarbeit A_v

Bruchzähigkeit K_{IC}

Wöhlerkurve

Dauerfestigkeit
Zeitfestigkeit

Korrosions – beständigkeit

Gefüge

Textur (O D F)

Kornformorientierung (Or. – Korrelation)

Zweitphasen

Einformung , Orientierung

Bild 2. Werkstoffeigenschaften vor der Umformung

Zur Ermittlung dieser Eigenschaften dient vor allem der Zugversuch /6 bis 15/, der auch in der Normung eine bevorzugte Rolle spielt. Es ist aber zu bedenken, daß das Umformvermögen metallischer Werkstoffe bei hydrostatischer Druckspannung größer ist als bei hydrostatischer Zugspannung /16, 17/, vgl. Bild 3. Deshalb wird der Stauchversuch /18 bis 20/ bevorzugt, wenn die Fließkurve bis zu hohen Umformgraden bestimmt werden soll, was zum Verständnis mancher Umformverfahren erforderlich ist. Der Torsionsversuch /21, 22/ ist vergleichsweise wenig verbreitet, weil er eine spezielle Versuchseinrichtung erfordert, während Zug- und Stauchversuch auf den meisten handelsüblichen Zerreißmaschinen durchgeführt werden können.

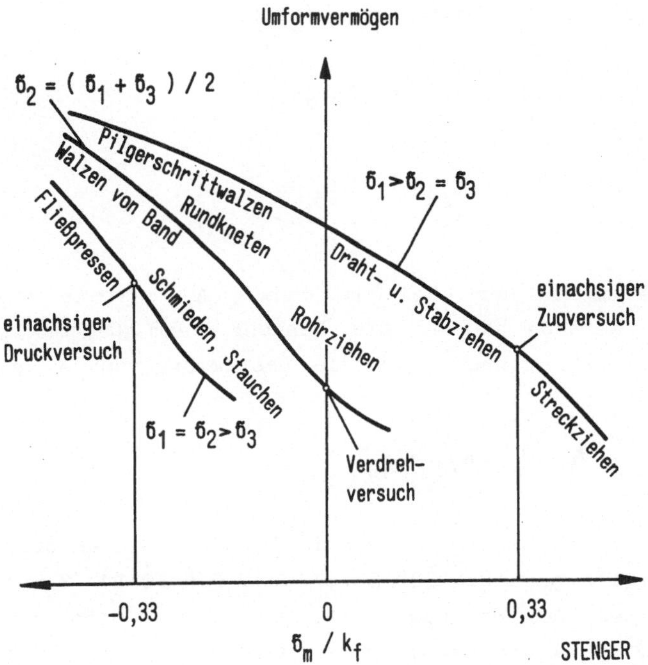

Bild 3: Umformvermögen metallischer Werkstoffe in Abhängigkeit vom hydrostatischen Spannungsanteil /16/.

Für die Aufnahme der Fließkurven dünner Bleche im Bereich höherer Umformgrade gibt es einige spezielle Versuche wie den hydraulischen Tiefungsversuch /23 bis 25/ und den ebenen Torsionsversuch /26, 27/, die noch nicht völlig ausgereift sind. Es sei auch darauf hingewiesen, daß bei Blechwerkstoffen vielfach die sog. "Hollomon-Gleichung" (auch "Ludwik-Gleichung" genannt) vorausgesetzt werden kann:

$$k_f (\varphi) = C \varphi^n \tag{1}$$

Hierin sind C und n werkstoffabhängige Konstante (n = "Verfestigungsexponent"). In diesem Falle genügt es, die beiden Konstanten C und n mit den im Flachzugversuch ermittelten Kennwerten Gleichmaßdehnung und Zugfestigkeit zu verknüpfen. Es gelten die Beziehungen /14/:

$$n \approx \varphi_g = \ln (1+A_g) \tag{2}$$

und

$$C \approx R_m \left(\frac{e}{n}\right)^n \tag{3}$$

Zur Bestimmung der Gleichmaßdehnung A_g besteht die einfachste Methode darin, die Dehnung beim Lastmaximum zu messen. Etwas genauer ist die Bestimmung von A_g aus der Beziehung

$$A_g = 2A_{10} - A_5 \tag{4}$$

Hierin sind A_{10} und A_5 die Bruchdehnungen, die an proportionalen Proben nach DIN 50125 /6/ bzw. DIN 50114 /8/ gemessen wurden; es ist vorausgesetzt, daß die Einschnürzone in beiden Probentypen die gleiche Länge hat. Dieses Verfahren ist auch noch nicht sehr genau, da schon kleine Maßabweichungen der Probe sich empfindlich auf die gemessene Bruchdehnung auswirken /15/. Deshalb ist es in der Blechumformung üblich, den n-Wert aus zwei Punkten der gemessenen Last-Weg-Kurve zu bestimmen /28, 29/. Aus Gl. (1) kann die

Beziehung abgeleitet werden

$$n = \frac{\ln \dfrac{(1+\varepsilon_2)}{(1+\varepsilon_1)} + \ln\left(\dfrac{F_2}{F_1}\right)}{\ln \dfrac{\ln(1+\varepsilon_2)}{\ln(1+\varepsilon_1)}} \tag{5}$$

Hierin sind F_1 und F_2 die gemessenen Kraftwerte, bei den Dehnungen und mit $= L/L_0$ (für und werden Werte unterhalb der Gleichmaßdehnung gewählt).

Vergleichende Betrachtungen der Methoden zur Aufnahme von Fließkurven finden sich in /30 bis 36/, für den Sonderfall dünner Bleche in /37/.

Bei Blechen ist zur Beurteilung der Umformeignung i. allg. auch die Bestimmung der Anisotropieeigenschaften erforderlich, sei es durch Ermittlung des r-Wertes im Flachzugversuch /38/ oder durch die aufwendigere Ermittlung der Fließortkurve /39/.

Zur Erfassung des Umformvermögens dient im Bereich der Massivumformung neben den im Zugversuch zu bestimmenden Kennwerten Bruchdehnung und Brucheinschnürung auch die Stauchbarkeit (upsettability /40, 41/, s. a. Abschnitt 4.2), oder der Drehwinkel bis zum Bruchbeginn im Torsionsversuch /42/, ferner bei Blechwerkstoffen die Erichsen-Tiefung /43, 44/ oder auch die Kerbzugdehnung /45, 46/. Die Bedeutung der Brucheinschnürung für die Blechumformung wird eindrucksvoll demonstriert durch die Beobachtung, daß Bleche gerade dann eine scharfkantige 180°-Biegung ertragen, wenn die Brucheinschnürung mindestens 50% beträgt /47/.

Für die Praxis ist die sog. "Grenzformänderung" oft wichtiger als das Umformvermögen. Die Grenzformänderung ist keine reine Werkstoffeigenschaft, sondern eine Eigenschaft des Systems Werkstück-Werkzeug unter Berücksichtigung der Reibung. Die Aufnahme von Grenzformänderungskurven hat vorwiegend im Bereich der Blechumformung Bedeutung erlangt /48 bis 52/.

2.3   Oberflächeneigenschaften

Die Aufzählung der Versuche zur Beurteilung der Umformeignung wäre unvollständig, wenn nicht auch die Ermittlung der Randschicht- und Oberflächeneigenschaften berücksichtigt würde. Diese Thematik wird aber auch in anderen Tagungsbeiträgen behandelt, so daß hier einige pauschale Hinweise genügen sollen.

Die Oberfläche wird einerseits charakterisiert durch ihre Mikrogeometrie, d.h. vor allem durch Rauheitskennwerte /53 bis 60/. Diese Charakterisierung ist mit einigen Problemen verbunden, was sich darin äußert, daß nicht weniger als 12 DIN-Normen diesem Thema gewidmet sind. Von diesen sind allerdings nur einige /53 bis 57/ von unmittelbarer Bedeutung für die Umformtechnik.

Zur Charakterisierung der Randschicht bzw. Oberfläche gehören andererseits ihre chemischen Eigenschaften, die z.T. durch recht aufwendige Analysenmethoden bestimmt werden /61 bis 63/, sowie ihre physikalischen und mechanischen Eigenschaften /64 bis 67/.

2.4   Weitere Eigenschaften des nicht umgeformten Werkstoffes

Die in Bild 2 rechts oben zusammengefaßten Eigenschaften dienen nicht unmittelbar der Beurteilung der Umformeignung (allerdings korrelieren manche dieser Eigenschaften mit einigen der zuvor genannten Kennwerte).

Es sind dies technologische Kennwerte wie die Härte /63 bis 70/ und die Kerbschlagarbeit /71/, Bruchmechanik-Kennwerte wie die Bruchzähigkeit k_{Ic} /72/, ferner das Ermüdungsverhalten /73, 74/, die Zeitfestigkeit /75, 76/ und die Korrosionsbeständigkeit /77 bis 83/. Diese Eigenschaften sind schon beim nicht umgeformten Werkstoff von Interesse, da sie möglicherweise eine Aussage über zu erwartende Eigenschaften des umgeformten Werkstückes gestatten.

Zur Ermittlung der genannten Eigenschaften dienen Prüfverfahren, die größtenteils in der DIN genormt sind, in Einzelfällen werden ASTM-Standards herangezogen /72, 83, 84/.

Die Eigenschaften der dritten Gruppe, die in Bild 2 unten zusammengefaßt sind, und das Gefüge mit den dazugehörenden Merkmalen betreffen, sind von Bedeutung, wenn versucht wird, Eigenschaften der ersten oder zweiten Gruppe metallkundlich zu deuten oder gezielt zu verändern.

Der Erfassung des Gefüges[*] dient als klassische Methode die lichtoptische metallographische Gefügeuntersuchung /84 bis 92/, die ggf. durch aufwendige moderne Prüfverfahren, wie z.B. die Röntgentexturanalyse /93, 94/ und die Elektronenmikroskopie /95, 96/ ergänzt werden kann.

Zusammenfassend kann über die Möglichkeit, Werkstoffe anhand von Versuchen vor der Umformung zu beurteilen, folgendes ausgesagt werden:

1. Versuche am Werkstoff vor der Umformung haben aus der Sicht der Umformtechnik einen Vorrang vor der theoretischen Berechnung von Werkstoffeigenschaften. Niemand wird z.B. auf die Idee kommen, eine Fließkurve für Anwendungen in der

---

[*] Zu den Gefügeeigenschaften im weiteren Sinne können auch die Eigenspannungen 2. Art gezählt werden. Diese spielen aber vor der Umformung (bei weichgeglühtem Material) keine wesentliche Rolle und werden daher erst bei der Betrachtung des umgeformten Werkstückes in Abschnitt 4 berücksichtigt.

Umformtechnik theoretisch zu berechnen. Allenfalls kann aus der chemischen Analyse des Werkstoffes zusammen mit bekannten mechanischen Kennwerten die Fließkurve näherungsweise bestimmt werden /97, 98/. Dies ist aber keine theoretische Bestimmung der Fließkurve, sondern nur eine spezielle Art empirischer oder halbempirischer Vorgehensweise.

Im Prinzip können Fließkurven auch aus der Literatur entnommen werden. Seit einigen Jahren liegen verschiedene Fließkurvensammlungen vor /99 bis 102/. Da aber die Fließkurve eines gegebenen Werkstoffes vom jeweiligen Gefügezustand abhängt, kann in den meisten Fällen auf eine experimentelle Bestimmung der Fließkurve nicht verzichtet werden.

Auch andere Werkstoffkennwerte sind in Datensammlungen zusammengestellt, s. z. B. /103 bis 105/, jedoch gilt wie im Fall der Fließkurve, daß eigene Versuche in vielen Fällen unverzichtbar sind.

2. Es bestehen charakteristische Unterschiede in Bezug auf die geforderte Information und in Bezug auf die in Frage kommenden Prüfverfahren zwischen Probematerial für die Massivumformung einerseits und dünnen Blechen andererseits (wobei als obere Grenze für die Blechdicke etwa 3 mm anzusetzen ist):
- Für die Massivumformung interessiert in vielen Fällen das Werkstoffverhalten nicht nur bei Raumtemperatur, sondern auch bei erhöhter Temperatur, während dünne Bleche meistens nur kalt umgeformt werden. Außerdem werden in einigen Verfahren der Massivumformung höhere Umformgrade erreicht als in der Blechumformung. Dagegen spielen bei Blechwerkstoffen die Anisotropieeigenschaften eine größere Rolle als bei massivem Probematerial.
- Die zur Verfügung stehenden Prüfverfahren sind im Fall massiven Probematerials vielfältiger als bei Blechwerkstoffen.
- Bei der Blechumformung ist in der Regel die Bedeutung von Oberflächenphänomenen größer als bei der Massivumformung.

3. Für niedrige Umformgrade und Umformgeschwindigkeiten und
für Raumtemperatur sind die in Frage kommenden Versuche zur
Bestimmung der Umformeignung vielfältiger als für hohe Umformgrade, hohe Umformgeschwindigkeiten und erhöhte Temperaturen.

4. Probleme durch eine Forderung nach scharfer räumlicher
Auflösung bestehen außer bei Oberflächen- und Randschichtun-
tersuchungen in der Regel nicht, da das Probematerial
meistens als (mehr oder weniger) homogenes Halbzeug vorliegt. Dies ist ein markanter Unterschied zu den im folgenden betrachteten Versuchen während und nach der Umformung
und weist darauf hin, daß Versuche zur Erfassung von Werkstoffeigenschaften vor der Umformung vielfach genauere Ergebnisse liefern als Versuche während und nach der Umformung.

3 Der Umformvorgang

3.1 Allgemeines

Beim Umformvorgang wird - von exotischen Umformvorgängen ab-
gesehen - das Werkstück durch eine mechanisch aufgebrachte
Kraft bzw. Kräfte und die dadurch im Werkzeug hervorgerufe-
nen Auflagerkräfte umgeformt. Daher kann die "Wechselwirkung zwischen Werkstoff und Umformung" gewissermaßen in der
Wirkfuge zwischen Werkstück und Werkzeug lokalisiert gedacht werden. Aus schon erwähnten Gründen wird jedoch hier
auf die Vorgänge in der Wirkfuge nicht eingegangen.

Es werden nur wenige ausgewählte Versuche betrachtet, in
denen während des Umformvorganges Informationen gewonnen
werden.

Die Messung der Umformkraft erfolgt i. allg. mit einem im
Kraftfluß eingebauten Kraftmeßkörper, dessen elastische Verformung in eine elektrische Größe umgewandelt und dann

verstärkt wird, s. z.B. /106/. Die Umformkraft wird
meistens in Abhängigkeit vom Werkzeugweg (Umformweg) gemessen, vgl. Bild 4. Zur Messung dienen induktive Wegaufnehmer.

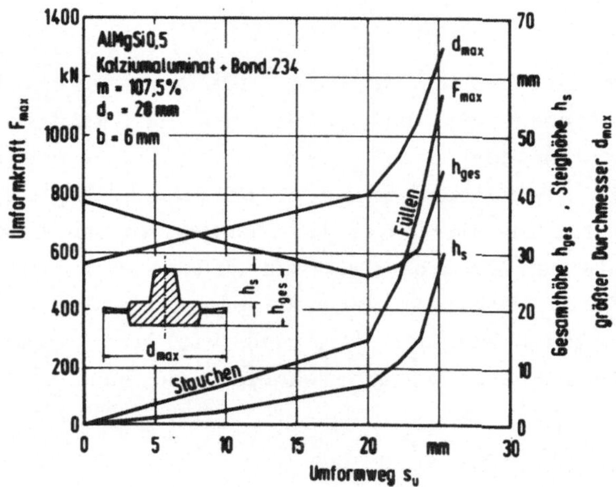

Bild 4. Beim Kaltgesenkschmieden gemessener Kraft-Weg-Verlauf
/1o7/.

3.2 Messungen der Kontaktnormalspannungen in der Wirkfuge

Die Messung der in der Wirkfuge zwischen Werkzeug und
Werkstück senkrecht zur Werkstückoberfläche wirkenden Kon-
taktnormalspannungen ist von Bedeutung für die Auslegung
der Werkzeuge, für plastizitätstheoretische Berechnungen
der Umformvorgänge wie auch für eine Behandlung von Fragen
der äußeren Reibung, der Maßgenauigkeit und der Werkzeug-
standzeit (eine theoretische Berechnung der Spannungen ist
oft nicht ausreichend genau).

Es sind verschiedene Methoden zur Messung der Kontaktnormal-
spannungen bekannt /107 bis 110/. In /107/ wurden zwei der

wichtigsten experimentell erprobt und miteinander verglichen: das Meßverfahren mit Meßstiften und das mit Meßscheiben (Sensoren).

Das Meßverfahren mit <u>Meßstiften</u>, vgl. Bild 5, beruht auf der Umwandlung der <u>elastischen</u> Verformung eines Meßkörpers in eine elektrisch meßbare Größe. Wird das Werkzeug durch die Kraft F belastet, so entfällt von dieser auf den örtlich vorhandenen Meßstift eine Teilkraft F_1. Zur Kalibrierung wird der Meßstift mit einer bekannten Einzelkraft oder einem bekannten hydraulischen Druck belastet.

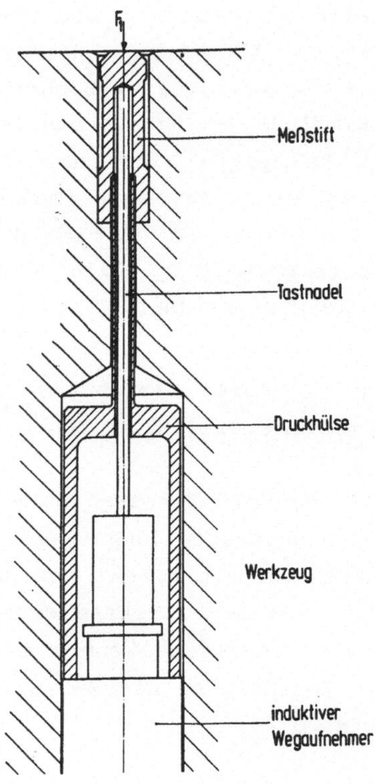

Bild 5. Schematischer Aufbau der Meßstifte zum Ermitteln der Kontaktnormalspannung /1o7/.

Das Meßverfahren mit <u>Meßscheiben</u> beruht im Gegensatz zur Verwendung von Meßstiften auf der Messung <u>plastischer</u> Formänderungen. Eine dünne Metallscheibe (Sensor) mit einseitigem spitzkämmigem Profil (Bild 6) wird zwischen Werkstück und Werkzeug mit der Profilseite zum Werkzeug angebracht. Als Maß für die Kontaktnormalspannung dient die plastische Abplattung b der Spitzen.

Um den Umformvorgang möglichst wenig zu stören, ist die Dicke der Sensoren auf etwa 0,6 mm zu begrenzen /105/. Der Durchmesser der Sensoren beträgt zwischen 5 und 10 mm. Der Zusammenhang zwischen Abplattung b und Kontaktnormalspannung wird durch eine Kalibrierung bestimmt. Bild 6 zeigt eine Kalibrierkurve für X5CrNi18 9. Nach /107/ ist es notwendig, für jeden Sensorwerkstoff gesondert zu kalibrieren; dagegen ist der Einfluß des Schmierstoffes auf die Kalibrierkurve gering. Es zeigte sich, daß die Kalibrierkurve für C45 nur bis zu einer Kontaktnormalspannung von etwa 1400 N/mm^2 linear verläuft. Im Bereich höherer Spannungen ist daher der Sensorwerkstoff X5CrNi18 9 wegen des größeren linearen Bereiches besser geeignet.

Zur Herstellung der Sensoren bietet sich neben dem Zerspanen die Fertigung durch Profilwalzen an.

In /107/ wurden aus der mit Sensoren beim Kaltgesenkschmieden ermittelten Spannungsverteilung - vgl. Bild 7 - die Umformkräfte berechnet und mit direkt gemessenen Werten verglichen. Die direkt gemessenen Kräfte waren etwas höher als die berechneten, aber insgesamt konnte die Übereinstimmung der Ergebnisse als Bestätigung des Meßverfahrens mit Sensoren angesehen werden.

Zum Vergleich der Meßverfahren mit Meßstiften und mit Sensoren ist nach /107/ noch folgendes zu bemerken:

Der wesentlichste Nachteil des Verfahrens mit Meßstiften ist der hohe Aufwand für Meßvorrichtungs- und Werkzeugbau, wobei enge Toleranzen einzuhalten sind. Auch während der Versuchsdurchführung müssen die Meßstifte kontrolliert werden, da die Lage der Meßflächen zur Werkzeugoberfläche einen Einfluß auf das Meßergebnis hat. Sobald sich die Meßflächen im unbelasteten Zustand unterhalb der Werkzeugoberfläche befinden, müssen die Meßstifte ausgetauscht und neu kalibriert werden.

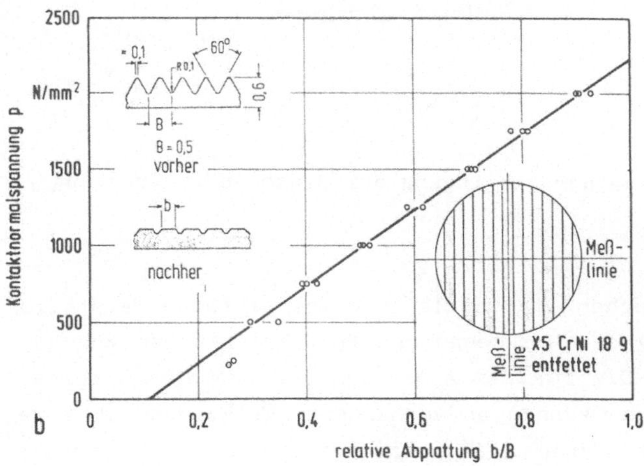

Bild 6. Sensoren zur Bestimmung der Kontaktnormalspannung und Kalibrierkurve /1o7/.

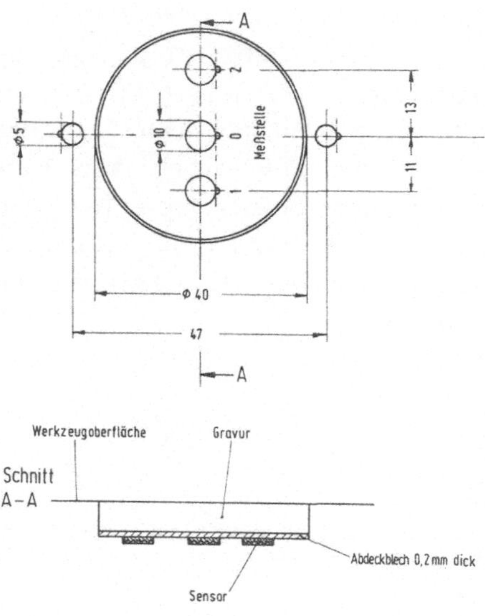

Bild 7. Lage und Anbringung der Sensoren im Werkzeug /1o7/

Daher sprechen die vergleichsweise einfache Herstellung und
Handhabung der Werkzeuge und Meßeinrichtungen zugunsten der
Sensoren. Die für das Anbringen der Sensoren erforderlichen
Vertiefungen können an vorhandenen Werkzeugen ohne weiteres
angebracht werden. Für das Verfahren mit Sensoren gilt
allerdings die Einschränkung, daß es nur zur Bestimmung von
Maximalwerten geeignet ist (allenfalls sind mit Stufenwer-
ten zusätzliche Ergebnisse zu gewinnen). Außerdem können
die Sensoren nur an Stellen mit geringem Werkstofffluß
eingesetzt werden, da sonst die Meßwerte zu sehr verfälscht
werden. Um diese Fehlermöglichkeit zu unterdrücken, ist es
möglich, die Sensoren mit dünnen Blechen abzudecken, vgl.
Bild 7. Nachteilig ist bei Sensoren auch der große Aufwand
für die Versuchsauswertung.

3.3 Stoffflußuntersuchungen und Formänderungsanalyse

Bei vielen Verfahren der Massivumformung ist es möglich, durch Teilen des Werkstückes und Aufbringen eines Liniennetzes, das während der Umformung verzerrt wird, das Geschwindigkeitsfeld näherungsweise zu bestimmen /111, 112/. Als Beispiel wird ein stationärer Vorgang, das axialsymmetrische Fließpressen, betrachtet. Bild 8 zeigt ein verzerrtes Liniennetz, das in einer Ebene, welche die Achse des Werkstückes enthält, angebracht wurde. Die ursprünglich zur Achse parallelen Linien des Gitters sind Strom- und Bahnlinien des Geschwindigkeitsfeldes.

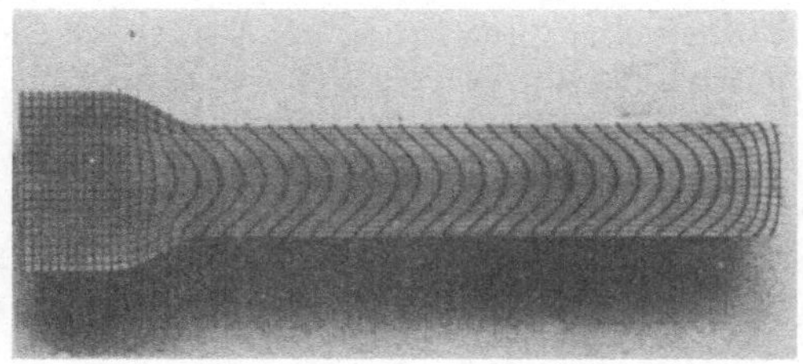

Bild 8. Visioplastische Stoffflußuntersuchung beim Vollvorwärtsfließpressen (Werkstoff Ck15, angefast, $\varphi_g \approx$ 0,9o, Schmierstoff MoS_2).

Auch bei instationären Vorgängen lassen sich die Geschwindigkeitsfelder auf ähnliche Weise ermitteln, wenn die Änderung der Gitterverzerrung für zwei dicht aufeinanderfolgende Schritte des Umformvorganges betrachtet wird.

Aus dem Geschwindigkeitsfeld können unter vereinfachenden Annahmen die Formänderungsgeschwindigkeiten errechnet werden, und aus diesen die Formänderungen selbst und damit - bei bekannter Fließkurve - auch die örtliche Verfestigung. Es ist aber auch möglich, die örtliche Formänderung über Härtemessungen zu bestimmen, vgl. Abschnitt 4.4.

Für den Fall der Blechumformung kann mit Hilfe aufgedruckter Liniennetze die Formänderungsverteilung ermittelt werden. Dabei ist ein Zerteilen des Werkstückes nicht erforderlich. Das Liniennetz - das hier zumeist aus kleinen Kreisen besteht - wird einfach auf die Blechoberfläche aufgedruckt /113/, s. a. /114/.

4 Eigenschaften von Werkstoff und Werkstück nach der Umformung

4.1 Überblick

Der Unterschied zwischen Bild 9 und der Darstellung der Eigenschaften vor der Umformung in Bild 2 ergibt sich einerseits daraus, daß nach einer - abgeschlossenen ! - Umformung die Umformeignung nicht mehr interessiert (daher haben z.B. die Zähigkeitskennwerte nicht die gleiche Bedeutung wie vor der Umformung). Eher ist noch die Eignung des umgeformten Werkstoffes für andere nachfolgende Fertigungsverfahren von Bedeutung. Dieses Thema wird in Abschnitt 4.2 am Beispiel der spanenden Bearbeitung behandelt.

Andererseits gilt, wenn nach der Umformung ein gebrauchsfertiges Werkstück vorliegt, so daß keine weiteren Fertigungsschritte folgen (es ist hier in erster Linie an die Stückgutfertigung gedacht), das folgende.

1. Das Werkstück ist i. allg. inhomogen umgeformt, woraus sich mit entsprechenden Formänderungsgradienten zugleich Eigenspannungen 1. Art ergeben (sofern nicht oberhalb der Rekristallisationstemperatur umgeformt wurde). Nur in. Ausnahmefällen liegt eine annähernd homogene Formänderung vor, so daß von echten Werkstoffeigenschaften frei von Geometrieeinflüssen gesprochen werden kann. Dieser in der Praxis in völliger Strenge nie gegebene Fall, der aber von grundsätzlichem Interesse ist, wird in Abschnitt 4.3 betrachtet.

Bild 9. Werkstoff- bzw. Werkstückeigenschaften nach der Umformung

2. Zur Beschreibung des Werkstückes mit Hilfe von Stoffeigenschaften finden sich einige Angaben in Abschnitt 4.4.

3. Im allgemeinen Fall einer inhomogenen Umformung interessieren Versuche zur Bestimmung von Eigenspannungen 1. Art. Diesem Thema ist der Abschnitt 4.5 gewidmet.

Allgemein besteht bei einer inhomogenen Formänderungsverteilung das Problem der endlichen räumlichen Auflösung der Meßverfahren. Diese Aussage muß allerdings relativiert werden: Das Problem der räumlichen Auflösung besteht vorwiegend dann, wenn das Werkstück durch Stoffeigenschaften beschrieben werden soll, analog zum unverformten Werkstoff. Diese Beschreibung des Werkstückes ist aber nicht ganz angemessen, denn es werden dabei Begriffe wie z.B. die Streckgrenze, die für einen Werkstoff - im Prinzip unabhängig von Geometrieeinflüssen - definiert sind, auf einen Gegenstand angewandt, der erst durch seine Geometrie wesentlich zu dem wird, was er ist.

4. Daher ist die Beschreibung des Werkstückes durch Stoffeigenschaften zu ergänzen durch die Angabe solcher Eigenschaften, die das Werkstück <u>als Ganzes</u> charakterisieren: Hierbei besteht naturgemäß kein Problem durch eine Forderung nach guter räumlicher Auflösung.

Strenggenommen gehören zwar Eigenschaften, die das Werkstück als Ganzes charakterisieren, nicht mehr zum Thema "Wechselwirkung zwischen Werkstoff und Umformung". Aber die vorangegangene Betrachtung der Stoffeigenschaften ist ja kein Selbstzweck, sondern zielt darauf ab, das Umformverfahren oder Eigenschaften des umgeformten Werkstückes zu optimieren. Daher sollen Eigenschaften, die das Werkstück als Ganzes beschreiben, nicht völlig ausgeklammert bleiben. In Abschnitt 4.6 wird auf sie kurz eingegangen. Es sind dies z.B. Größe, Gestalt und Maßhaltigkeit des Werkstückes, ferner Oberflächen- und Randschichteigenschaften, vor allem aber die Eignung für die spezielle vorgesehene Funktion des fertigen Werkstückes, das in diesem Zusammenhang zweckmäßig als Bauteil bezeichnet und als Teil des Systems, in dem es eine Funktion hat, betrachtet wird.

4.2   Zerspanbarkeit

Im Prinzip ist ein Zerspanungsvorgang weiter nichts als ein Umformvorgang mit nachfolgender Werkstofftrennung. Es ist aber in keiner Weise ausreichend, die Zerspanbarkeit anhand der Umformeignung zu beurteilen, zumal sich in der Praxis beide Eigenschaften meist gegenläufig verhalten: ein gut zerspanbarer Werkstoff ist meist schlecht umformbar, und umgekehrt.

Zur Definition der Zerspanbarkeit ist zu beachten, daß in der DIN derzeit nur der Entwurf einer Norm existiert, in welcher die Zerspanbarkeit wie folgt definiert ist /115/: "Zerspanbarkeit ist die Eigenschaft eines Werkstoffes, sich unter gegebenen Bedingungen spanend bearbeiten zu lassen". Konkrete Angaben zur experimentellen Bestimmung der Zerspanbarkeit enthalten dagegen die Stahl-Eisen-Prüfblätter des VDEh. Hierin werden Versuche genannt, in denen quantitative Meßgrößen zur Beurteilung der Zerspanbarkeit gewonnen werden, wie z.B. die Messung der "spezifischen" Schnittkraft /116/ oder die Spanbeurteilung /117/.

Es interessiert nun die Frage, wie sich eine Umformung auf das Zerspanungsverhalten eines Werkstoffes auswirkt. Diese Frage ist für die Praxis immer dann von Bedeutung, wenn eine Umformung mit nachfolgender spanender Bearbeitung notwendig ist. Für diese Fertigungsfolge kommen im Prinzip zwei Gruppen von Werkstoffen in Frage (wobei die Betrachtung hier auf eine Kalt- oder Halbwarmumformung mit nachfolgender Zerspanung beschränkt bleibt):

1. Werkstoffe, die im Hinblick auf die Umformeignung optimiert sind, wie speziell die Kaltfließpreßstähle (C15, 16MnCr5, 42CrMo4, 100Cr6 usw.). Bei diesen Werkstoffen ist die nachfolgende Zerspanung u.U. problematisch, da sich z.B. Wirrspäne bilden können.

2. Werkstoffe, die im Hinblick auf die Zerspanungseigen-
schaften optimiert sind, wie die Automatenstähle (9S20,
10S20, 35S20 usw.). Bei diesen relativ spröden Werkstoffen
kann die Umformung problematisch sein.

Um einen Vergleich der beiden Werkstoffgruppen im Hinblick
auf die Fertigungsfolge Kaltumformung-Zerspanung zu ermöglichen, wurden am Institut für Umformtechnik der Werkstoff
C15 als Vertreter der Kaltfließpreßstähle und 9SMn28, 10S20
sowie 35S20 als Vertreter der Automatenstähle experimentell
untersucht /118/.

Zum Vergleich der zwei genannten fertigungstechnischen Alternativen wurde sowohl das Umformverhalten der beiden Werkstoffgruppen als auch ihr Verhalten bei einer auf die
Umformung folgenden spanenden Bearbeitung verglichen. Die
Erwartung, daß der Automatenstahl im Hinblick auf die Zerspanung, und der Fließpreßstahl im Hinblick auf die Umformung überlegen ist, wurde durch die Versuchsergebnisse im
Großen und Ganzen bestätigt. Dies soll hier exemplarisch
aufgezeigt werden.

Als Beispiel für Versuche zur Erfassung der Umformeignung
ist in Bild 10 eine Probe zur Bestimmung der Stauchbarkeit
(Upsettability) dargestellt.

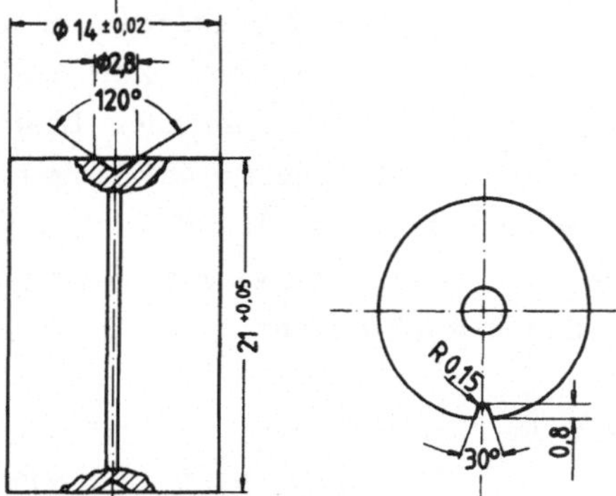

Bild 1o. Zylinderstauchproben mit Längsnut zur Ermittlung der
Stauchbarkeit (upsettability /41/).

In Bild 11 sind die damit erhaltenen Versuchsergebnisse
eingezeichnet. Die "kritische relative Höhenänderung", bei
der die Anrißbildung eine bestimmte Größe erreicht hat, als
Maß für die Stauchbarkeit ist bei Raumtemperatur für den
Fließpreßstahl C15 am größten (bei Halbwarmtemperatur allerdings für 10S20, während der C15 in diesem Fall erst an
zweiter Stelle folgt).

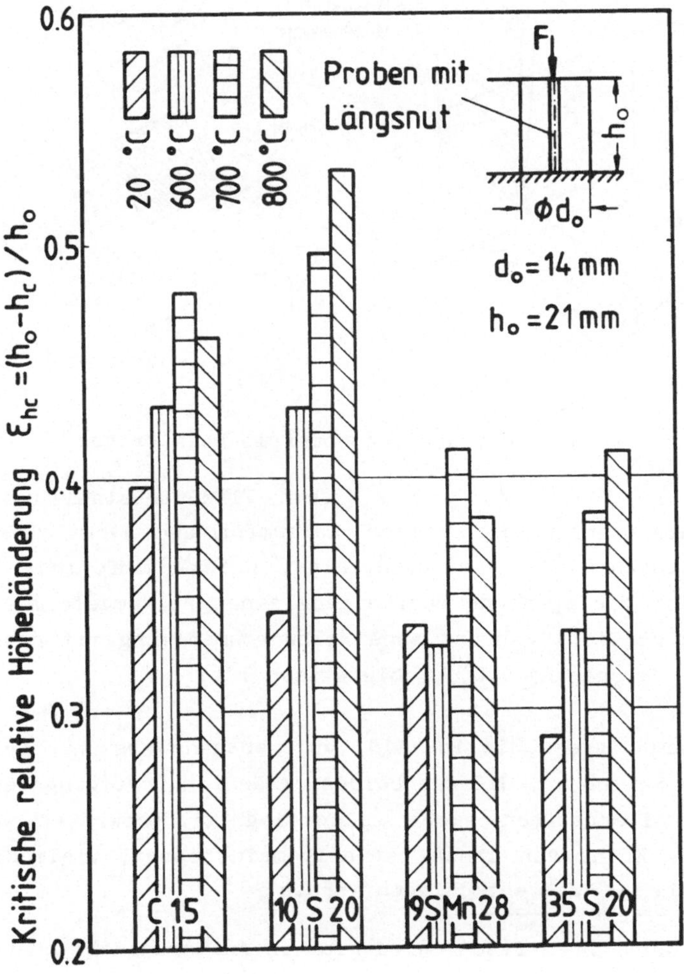

Bild 11. Prüfung der Stauchbarkeit: "kritische relative Höhenänderung" verschiedener Stähle /118/

Als Beispiel für eine quantitative Erfassung des Zerspanungsverhaltens diene die Messung der "spezifischen" Schnittkraft k_c nach /116/, vgl. Bild 12[*].

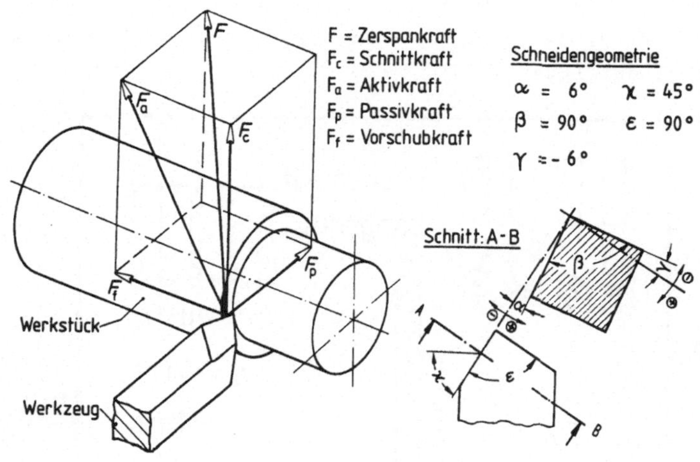

Bild 12. Kräfte und Schneidengeometrie beim Drehen

Wie Bild 13 zeigt, ist k_c beim Fließpreßstahl und den Automatenstählen nach einer Kaltumformung nicht deutlich verschieden, wohl aber nach einer Halbwarmumformung. Demnach ist die spanende Bearbeitung eines kaltumgeformten C15 leicht möglich, während eine Halbwarmumformung mit nachfolgender Zerspanung u.U. problematisch ist.

Insgesamt lassen die in /118/ angegebenen Versuchsergebnisse den Schluß zu, daß die beiden erwähnten fertigungstechnischen Alternativen nicht allgemeingültig bewertet werden können. Keine von ihnen ist allgemein besser, vielmehr muß von Fall zu Fall entschieden werden.

---

[*]Das Wort "spezifisch" wird hier in Anführungsstrichen geschrieben, weil die Schnittkraft auf den Spanungsquerschnitt bezogen ist, d.h. auf eine Fläche, während üblicherweise spezifische Größen auf eine Masse bezogen werden.

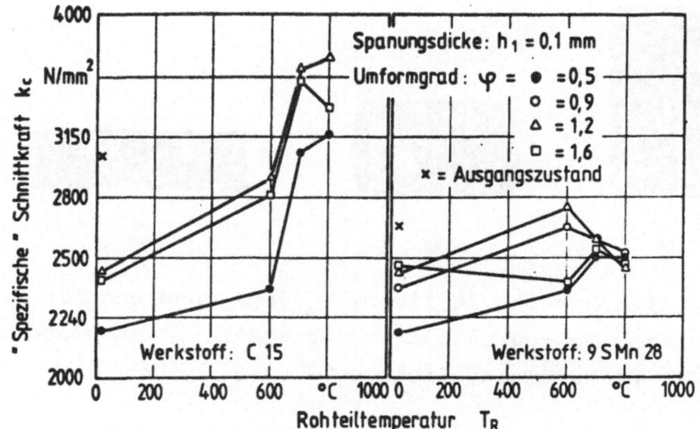

α	γ	λ	ε	κ	r_ε
6°	-6°	6°	90°	45°	0,8 mm

Bild 13. "Spezifische" Schnittkräfte umgeformter Stähle in Abhängigkeit von der Rohteiltemperatur /114/.

4.3 Homogene Umformung zur Ermittlung echter Werkstoffeigenschaften

Eine räumlich homogene Umformung läßt sich im Zugversuch bis zum Erreichen der Gleichmaßdehnung realisieren. Dies ist eine seit langem praktizierte Methode, um den Einfluß einer Umformung auf Werkstoffeigenschaften zu bestimmen. Durch Entnahme von Probematerial aus der gedehnten Zugprobe lassen sich im Prinzip alle Werkstoffeigenschaften in Abhängigkeit vom Umformgrad bestimmen. Allerdings ist der bei Gleichmaßdehnung erreichte Umformgrad für die meisten Werkstoffe relativ niedrig, vgl. Gl. (2).

Um höhere Umformgrade ohne Formänderungsgradienten zu erhalten, ist es nötig, einen anderen experimentellen Weg zu beschreiten. Eine - vielleicht die einzige praktikable - Möglichkeit hierzu bietet der Zylinderstauchversuch mit Schmierung nach Rastegaev /119 bis 121/, vgl. Bilder 14 und 15. Dieser ermöglicht es, Proben bis zu Umformgraden höher

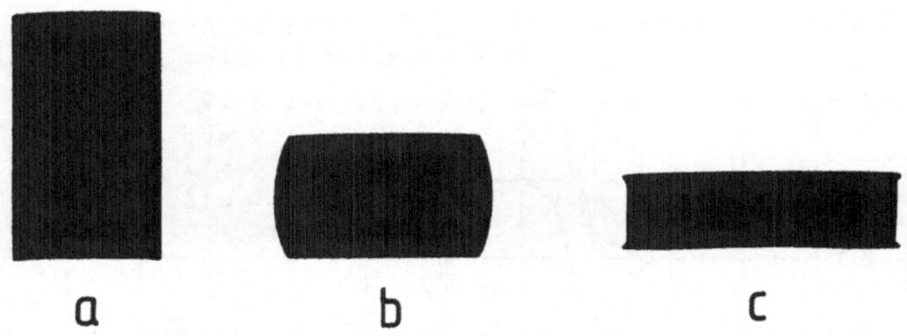

Bild 14. Stauchproben (h_o=16mm, r_o=5mm) ohne und mit Schmier-
taschen nach Rastegaev : a) ungestaucht, b) ohne Schmierung,
$\varphi \approx$ o,7, c) Rastegaev-Probe, $\varphi \approx$ 1,3 /121/.

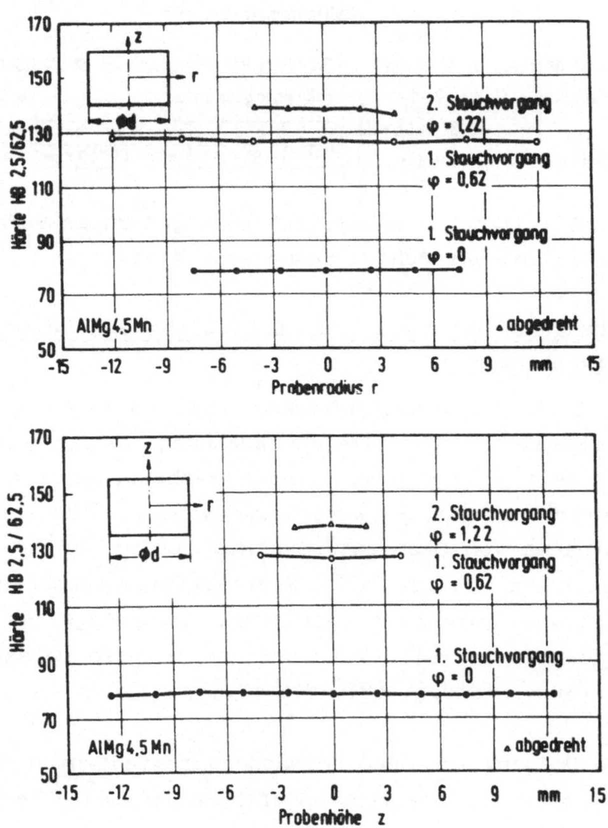

Bild 15. Härteverteilung über Querschnitt und Höhe gestauch-
ter Rastegaev-Proben /1o7/.

als $\varphi = 1$ homogen umzuformen und bietet damit die Voraussetzung, um - bei entsprechender Größe der gestauchten Proben - Eigenschaften eines bis zu hohen Umformgraden homogen umgeformten Werkstoffes experimentell zu bestimmen. Da diese Möglichkeit anscheinend noch nicht genutzt wurde, wird z.Z. am Institut für Umformtechnik eine Untersuchung durchgeführt, in der die Veränderung ausgewählter Werkstoffeigenschaften in Abhängigkeit vom Umformgrad bei homogener Stauchung nach Rastegaev bestimmt wird.

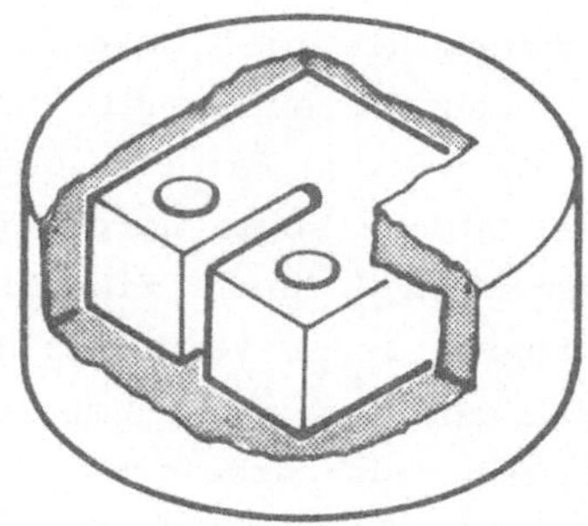

Bild 16. Entnahme einer CT-Probe zur Bestimmung von k_{Ic} nach ASTM E 399-81 aus einer gestauchten Rastegaev-Probe

Bild 16 zeigt als Beispiel, wie eine CT-Probe zur Bestimmung der Bruchzähigkeit (Rißzähigkeit) k_{Ic} nach ASTM-Standard /72/ aus einer gestauchten Rastegaev-Probe herausgearbeitet werden kann. Über Ergebnisse dieser Versuche soll zu einem späteren Zeitpunkt berichtet werden.

4.4 Bestimmung von Stoffeigenschaften des umgeformten Werkstückes

Wie erwähnt, genügt zur Beschreibung des fertigen Werkstückes die Angabe geometrieunabhängiger Stoffeigenschaften nicht. Trotzdem ist deren Bestimmung unverzichtbar, da die Gebrauchseigenschaften des Werkstückes ohne Kenntnis der Stoffeigenschaften nicht verstanden (oder gezielt verändert) werden können.

Die Bestimmung der Stoffeigenschaften erfolgt grundsätzlich in den gleichen Versuchen wie beim unverformten Werkstoff. Diese brauchen nicht nochmals aufgezählt zu werden. Allerdings ergeben sich in vielen Fällen, bedingt durch die endliche Größe der Werkstücke sowie durch Formänderungsgradienten und Eigenspannungen 1. Art, Einschränkungen in Bezug auf die Durchführbarkeit der Versuche: so ist es insbesondere oft nicht möglich, aus einem umgeformten Werkstück DIN-gerechte Zugproben herauszuarbeiten, weil die Stoffeigenschaften über die Länge eines Zugstabes veränderlich sind. Hier stellt sich die Forderung nach Versuchen mit möglichst guter räumlicher Auflösung.

Aus diesem Grunde wird manchmal, um beispielsweise Festigkeitskennwerte wie die Streckgrenze zu bestimmen, ein Umweg beschritten: es wird die räumliche Härteverteilung in dem (aufgesägten) Werkstück gemessen. Zusätzlich muß für den gegebenen Werkstoff der Zusammenhang zwischen Härte und Umformgrad bestimmt werden, wofür Rastegaev-Stauchproben besonders geeignet sind (vgl. Bild 15). Ist zudem die Fließkurve bekannt, so kann die Härte auch mit der Fließspannung bzw. Streckgrenze verknüpft werden, vgl. Bild 17.
Auf diese Weise wurde die in Bild 18 gezeigte Verteilung des Umformgrades und der Streckgrenze in einem umgeformten Werkstück ermittelt.

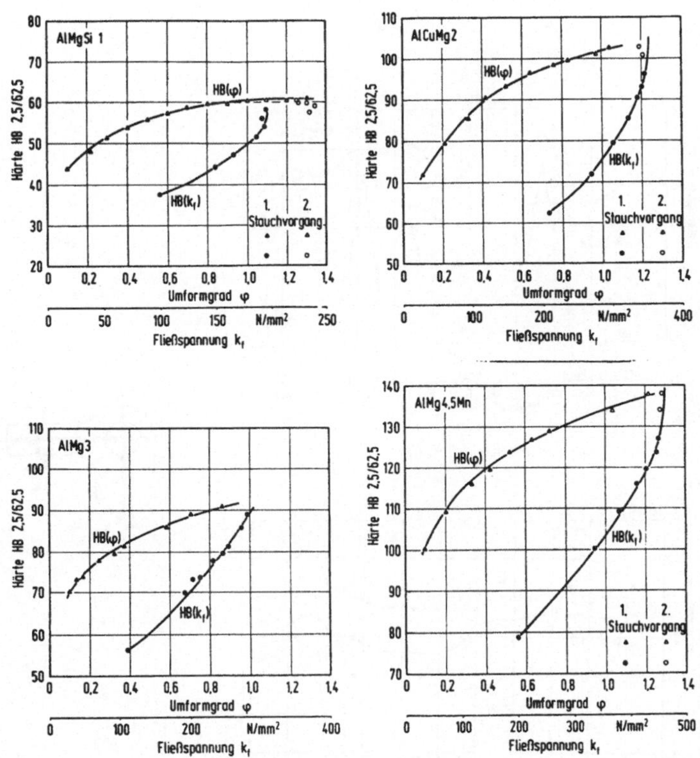

Bild 17. Härte von Rastegaev-Stauchproben im Zusammenhang mit Umformgrad und Fließspannung für einige Aluminiumlegierungen /1o7/ (für Stähle finden sich entsprechende Angaben in /122/).

In Bezug auf das Gefüge ist die Möglichkeit zu beachten, daß u.U. während der Umformung eine Phasenumwandlung erfolgt sein kann. Dies ist beispielsweise bei den nichtrostenden austenitischen Stählen von unmittelbarer praktischer Bedeutung, da hiervon sowohl die mechanischen Eigenschaften als auch z.B. die Korrosionsbeständigkeit beeinflußt werden. In diesem Fall ist eine Bestimmung des Martensitgehaltes nach einem magnetischen Verfahren möglich /123, 124/, es kann aber auch stattdessen röntgenographisch der Restaustenitgehalt ermittelt werden /125/.

Für das elastische Verhalten des Werkstückes und damit für seine mechanische Belastbarkeit ist neben der Streckgrenze

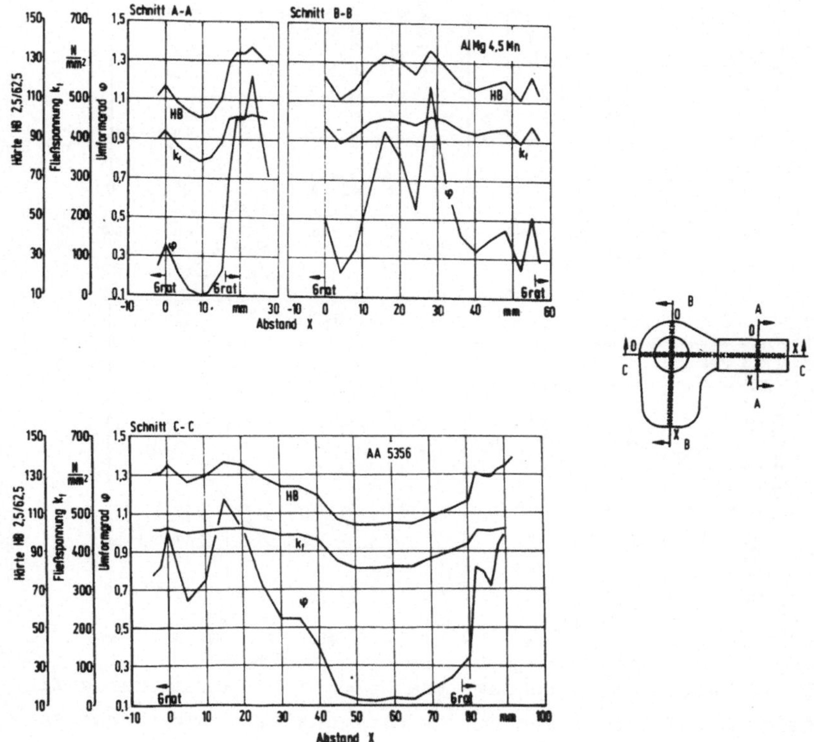

Bild 18. Verteilung von örtlicher Härte, Umformgrad und Fließspannung bzw. Streckgrenze bei einem Kaltgesenkschmiedeteil /1o7/

vor allem der E-Modul maßgeblich. Dieser kann durch eine Umformung nur insoweit verändert werden, als der E-Modul eines Einkristalles von der kristallographischen Richtung abhängt. Der E-Modul des vielkristallinen Werkstückes in einer gegebenen Richtung enthält die Orientierungsverteilungsfunktion (ODF) als Gewichtsfunktion sowie den von der kristallographischen Richtung abhängigen E-Modul des Einkristalles in einem Integral über alle kristallographischen Richtungen /126/.

Um den richtungsabhängigen E-Modul im Werkstück zu bestimmen, können - wenn die Abmessungen des Werkstücks dies zulassen - in unterschiedlichen Richtungen Zugproben entnommen werden. Dies ist aber sehr aufwendig und in der Praxis meist nicht erforderlich, da nur der E-Modul in Richtung der hauptsächlichen bzw. größten Beanspruchung interessiert.

4.5   Bestimmung von Eigenspannungen erster Art

Eine Folge der in der Praxis nach einer Kalt- oder Halbwarmumformung stets vorliegenden inhomogenen Formänderungsverteilung sind Eigenspannungen erster Art /127/. Während Mikroeigenspannungen (Eigenspannungen zweiter und dritter Art) selbst bei einer ideal homogenen Kalt- oder Halbwarmumformung auftreten - und deshalb als "Stoffeigenschaften" aufgefaßt werden können -, sind Eigenspannungen erster Art geometriebedingt. Besonders hohe Eigenspannungen erster Art sind zu erwarten, wenn der mittlere Umformgrad absolut hoch ist mit hohen räumlichen Gradienten, speziell also bei der Kaltmassivumformung.

Eigenspannungen - gleich ob Makro- oder Mikroeigenspannungen - sind nicht immer schädlich. Ihre Auswirkung hängt ab von dem Vorzeichen, das sie an der Werkstückoberfläche haben: Druckeigenspannungen in der Oberfläche erhöhen, und Zugeigenspannungen erniedrigen z.B. die Dauerfestigkeit und die Beständigkeit gegen Spannungsrißkorrosion.

Im Prinzip können bei einer Kaltumformung entstandene schädliche Eigenspannungen durch eine Wärmebehandlung nach der Umformung beseitigt werden. Dabei wird aber auch die in vielen Fällen erwünschte Verfestigung des Werkstoffes abgebaut; zudem können Maßabweichungen entstehen. Deshalb kommt eine Wärmebehandlung oft nicht in Betracht. Es besteht somit ein praktisches Interesse daran, die nach einer Kaltumformung vorliegenden Eigenspannungen zu bestimmen.

Unter den Verfahren zur experimentellen Bestimmung von Eigenspannungen erster Art[*] haben sich die mechanische und die röntgenographische Methode durchgesetzt.

Die mechanische Methode /128, 129/ beruht darauf, daß in einem spannungsbehafteten Körper durch spanende Bearbeitung o. ä. das Kräftegleichgewicht verändert wird. Dies führt zu einer meßbaren elastischen Verformung, aus der im Prinzip die Eigenspannungsverteilung berechnet werden kann. Dieses Verfahren hat allerdings eine schlechte räumliche Auflösung. Demgegenüber gestattet die röntgenographische Methode /128, 130, 131/ die Bestimmung der Eigenspannungen in einer Oberflächenschicht von einigen μ Dicke. Dabei beträgt der Durchmesser des Röntgenstrahles etwa 1 mm. Wenn nur die Eigenspannungen in der Mantelfläche des Werkstückes bestimmt werden sollen (diese sind für das Verhalten bei Ermüdung sowie bei Spannungsrißkorrosion maßgebend), ist die Röntgenmethode der mechanischen Methode überlegen, zumal sie zerstörungsfrei ist.

Es kann auch durch schichtweises Abtragen der Außenfaser die Eigenspannungsverteilung im Inneren des Werkstückes röntgenographisch ermittelt werden. Hierbei gehen allerdings in die Versuchsauswertung die gleichen Annahmen ein, die bei Anwendung der mechanischen Methode gemacht werden. Daher liefert die Röntgenmethode keine wesentlich über die mechanische Methode hinausgehende Information in Bezug auf die Spannungsverteilung über den Querschnitt des Werkstückes.

---

[*] Die experimentellen Möglichkeiten zur Bestimmung von Eigenspannungen zweiter und dritter Art sind beschränkt. Eigenspannungen zweiter Art können im Prinzip mit Hilfe der Röntgenbeugung abgeschätzt werden, während zur erperimentellen Bestimmung von Eigenspannungen dritter Art bisher kein für die Praxis brauchbares Verfahren zur Verfügung zu stehen scheint.

Die Röntgenbeugungsmessungen für sich allein sind auch in manchen Fällen schwierig zu interpretieren, da z.B. Textureinflüsse auftreten können.

Aus diesen Gründen scheint die Bestimmung von Umformeigenspannungen am zuverlässigsten zu sein, wenn Meßergebnisse sinnvoll mit einer theoretischen Berechnung verknüpft werden. Die theoretische Berechnung von Umformeigenspannungen hat in den letzten Jahren große Fortschritte gemacht, im wesentlichen bedingt durch die Weiterentwicklung der Finite-Element-Methode /132 bis 134/. Es ist anzunehmen, daß in Zukunft durch Verknüpfung berechneter und experimentell bestimmter Eigenspannungswerte recht zuverlässige Ergebnisse gewonnen werden.

4.6 Bestimmung der Gebrauchseigenschaften des umgeformten Werkstückes

Diese Problematik überschreitet teilweise das Thema "Wechselwirkung zwischen Werkstoff und Umformung", soll aber dennoch wegen ihrer Bedeutung für die Praxis kurz erörtert werden.

Neben den Stoffeigenschaften sind (vgl. Bild 9 rechts) Größe und Gestalt des umgeformten Werkstückes maßgeblich für seine Funktion.

Der Einfluß der Größe wird in erster Näherung durch die triviale Tatsache beschrieben, daß - für eine gegebene Form - z.B. bei statischer Beanspruchung die ertragbare maximale Last proportional zum Querschnitt des Werkstückes ist. Das Wort "Größeneinfluß" im engeren Sinne wird aber für Effekte zweiter Ordnung gebraucht. Diese sind, soweit der Werkstoff durch die Kontinuumsmechanik beschrieben werden kann, Gegenstand der Ähnlichkeitsmechanik /135, 136/. Daneben besteht auch ein metallurgischer Größeneinfluß, s. z.B. /137/. Versuchsergebnisse sind daher i. allg. größenabhängig, so daß im Laborversuch Proben mit etwa der gleichen

Größe wie im Praxisfall verwendet werden sollen.

Größe und Gestalt des Werkstückes sollen einer vorgegebenen Fertigungstoleranz genügen. In der Kaltumformung werden enge Toleranzen gefordert, speziell dann, wenn keine spanende Nachbearbeitung des Werkstückes erfolgt. Die Maßhaltigkeit eines umgeformten Werkstückes kann beeinträchtigt werden durch elastische Verformung des Werkzeuges sowie durch Wärmeausdehnung von Werkstück und Werkzeug bei der Umformung /138/, ferner durch nach der Umformung im Werkstück verbleibende Eigenspannungen.

Die in Bild 8 als nächstes angegebenen Oberflächen- und Randschichteigenschaften können im Rahmen dieser Betrachtung, wie schon angedeutet, nicht den Platz einnehmen, der ihnen eigentlich zukommt. Sie sind aber implizit mit zu berücksichtigen - und sogar äußerst wichtig -, wenn im folgenden das Verhalten des umgeformten Werkstückes bei betrieblicher Beanspruchung am Beispiel des Ermüdungsverhaltens und seine Untersuchung in Laborversuchen betrachtet wird.

Das Ermüdungsverhalten des umgeformten Werkstückes kann im Prinzip als Zusammenwirken der Einflüsse des Werkstoffes (einschließlich einer homogenen Umformung und der Oberflächeneinflüsse) sowie der Geometrie (Größe, Gestalt und Inhomogenitäten der Umformung) beschrieben werden.

Über die Werkstoffeinflüsse liegt eine umfangreiche Literatur vor, s. z.B. /139 bis 141/. Der Einfluß einer homogenen Umformung auf das Ermüdungsverhalten wurde bislang hauptsächlich am Beispiel des Zugversuches im Bereich der Gleichmaßdehnung untersucht /142/. Je nach dem Umformgrad kann sich eine Erhöhung oder Erniedrigung der Dauerfestigkeit ergeben /143/.

Der Effekt einer gezielten Oberflächenbeeinflussung auf das Dauerschwingverhalten wird in /144/ beschrieben. Druckeigenspannungen an der Oberfläche erhöhen i. allg.—wie schon erwähnt—die Dauerfestigkeit /139, 140, 145/. Der starke Einfluß der Oberflächenbeschaffenheit auf das Ermüdungsverhalten äußert sich auch darin, daß die Dauerfestigkeit einer Werkstoffprobe abnimmt, wenn die Oberfläche beschädigt wird.

Der Einfluß der Probengröße auf das Ermüdungsverhalten wird in /140/ behandelt, der Einfluß der Gestalt in /140, 141, 145/.

Es ist aber schwierig oder unmöglich, aufgrund dieser Kenntnisse das Ermüdungsverhalten eines Werkstückes im konkreten Einzelfall - gegeben durch Werkstoff, Umformverfahren, Werkstückgeometrie usw. - vorherzusagen. Hier sind Versuche am gegebenen Werkstück erforderlich, um zu gesicherten, praxisrelevanten Aussagen zu kommen.

Einige Angaben über das Ermüdungsverhalten umgeformter Werkstücke finden sich für das Umformverfahren Gesenkschmieden in /146 bis 148/. Für das Fließpressen scheinen bisher keine derartigen Angaben vorzuliegen.

Um das Ermüdungsverhalten umgeformter Werkstücke in der Praxis realistisch beurteilen zu können, genügen Dauerschwingversuche mit konstanter Amplitude und Frequenz nicht immer, da hierdurch die Betriebsbeanspruchung oft nicht gut simuliert wird. Es wäre demnach eine Untersuchung der "Betriebsfestigkeit" erforderlich - wobei eigentlich nicht das umgeformte Werkstück für sich allein, sondern die ganze Konstruktion, in welche es eingebaut wird, zu prüfen wäre. So lange aber an manchen umgeformten Werkstücken noch nicht einmal die einfachsten Dauerschwingversuche durchgeführt worden sind, könnten Betriebsfestigkeitsuntersuchungen dazu führen, daß der zweite Schritt vor dem ersten gemacht wird.

Wegen des hohen Aufwandes bei herkömmlichen Dauerschwingversuchen besteht ein Interesse an Kurzprüfverfahren, die

schon bei niedriger Lastwechselzahl eine Aussage über das
Werkstoff- oder Bauteilverhalten gestatten. Eine derartige
Möglichkeit bietet die Thermometrie /149, 150/. Dabei wird
die Tatsache ausgenutzt, daß jede plastische Umformung oder
Verformung zu einer Temperaturerhöhung führt. Da dem Ermü-
dungsbruch stets eine plastische Verformung bzw. Umformung
vorausgeht, kann die Lebensdauer über Temperaturmessungen
bestimmt werden. Diese Möglichkeit scheint bisher noch
nicht zur Untersuchung des Ermüdungsverhaltens umgeformter
Werkstücke genutzt worden zu sein.

Eine andere Möglichkeit, durch Versuche mit relativ niedriger Lastwechselzahl eine Information über das Ermüdungsverhalten umgeformter Werkstücke zu gewinnen, besteht darin,
die Wechselverformungskurve zu bestimmen. Da auch hierüber
in der Literatur noch keine Angaben vorzuliegen scheinen,
wird dieses Thema in einem z.Z. am Institut für Umformtechnik laufenden Forschungsvorhaben untersucht.

Zusammenfassend kann der Stand der' Dauerschwinguntersuchun-
gen umgeformter Werkstücke bzw. Bauteile wie folgt charakterisiert werden:
- Die Vorhersage der Dauerfestigkeit aufgrund bekannter
Gesetzmäßigkeiten wird umso unsicherer sein, je stärker bei
der Umformung die Oberflächeneigenschaften verändert werden.

- Die Versuche müssen am jeweiligen konkreten Werkstück
durchgeführt werden unter Berücksichtigung aller praxisrelevanten Einflußgrößen.

- Die Versuche sind aufwendig, zeit- und kapitalintensiv,
so daß die Anwendung eines Kurzprüfverfahrens zu wünschen
wäre.

5   <u>Schlußbemerkungen</u>

Die Erfassung der wechselseitigen Beeinflussung eines Werkstoffes mit einem Umformverfahren durch Versuche bereitet umso mehr Schwierigkeiten, je komplexer die Gesamtheit der zu berücksichtigenden Einflußfaktoren ist. Es ist grundsätzlich schwieriger, durch Versuche am umgeformten Werkstück zu verallgemeinerungsfähigen Aussagen zu gelangen, als durch Versuche am Werkstoff vor der Umformung. Daher löst sich die Untersuchung der Gebrauchseigenschaften umgeformter Werkstücke in eine Vielzahl von Einzelfällen auf.

Schrifttum

/1/   Lange,K.: The investigation of metal forming processes as
      part of a technical system, in: Advances in Machine Tool
      Design and Research, Proc. 1oth Int MTDR Conf., Manchester
      September 1969, S. 485-498, Pergamon Press, Oxford/New
      York 197o

/2/   Lange,K., Wilhelm,H.: Interaction between work-materials
      and forming-processes. Annals of the CIRP 25 (1976),
      S. 531-537

/3/   V.Bertallanfy,L.: General System Theory. Penguin, London
      1971

/4/   Ropohl,G.: Eine Systemtheorie der Technik. Hanser, Mün-
      chen/Wien 1979

/5/   Czichos, H.: Tribology. Elsevier, Amsterdam/Oxford/New
      York 1978

/6/   DIN 5o125: Prüfung metallischer Werkstoffe, Zugproben,
      Richtlinien für die Herstellung, April 1951

/7/   DIN 5o145: Prüfung metallischer Werkstoffe, Zugversuch,
      Mai 1975

/8/   DIN 5o114: Zugversuch ohne Feindehnungsmessung an Blechen,
      Bändern oder Streifen mit einer Dicke unter 3mm, August
      1981

/9/   DIN 51221, Blatt 1: Zugprüfmaschinen, Allgemeine Anfor-
      derungen, August 196o

/1o/ DIN 51221, Blatt 2: Zugprüfmaschinen, Große Zugprüfma-
      schinen und Universalprüfmaschinen, August 196o

/11/ DIN 5121o: Prüfung metallischer Werkstoffe, Zugversuch ohne Feindehnungsmessungen an Drähten, August 1961

/12/ Siebel, E., Schwaigerer,S.: Zur Mechanik des Zugversuches, Arch. Eisenhüttenwes. 19 (1948), S. 145-152

/13/ Lange,G.: Vereinfachte Ermittlung der Fließkurve metallischer Werkstoffe im Zugversuch während der Einschnürung der Probe,Arch.Eisenhüttenwes. 45 (1974), S.8o9-812

/14/ Reihle,M.: Ein einfaches Verfahren zur Aufnahme der Fließkurve von Stahl bei Raumtemperatur, Arch. Eisenhüttenwes. 32 (1961), S.331-336

/15/ El-Magd,E.: Ermittlung der Fließkurven im Zugversuch, Arch.Eisenhüttenwes. 45 (1974), S.83-89

/16/ Stenger,H.: Über die Abhängigkeit des Formänderungsvermögens metallischer Werkstoffe vom Spannungszustand, Diss. TH Aachen 1965

/17/ Vater,M., Lienhart,A.: Abhängigkeit des Formänderungsvermögens metallischer Werkstoffe vom Spannungszustand bei unterschiedlich hoher Temperatur und Formänderungsgeschwindigkeit, Bänder Bleche Rohre 13 (1972), S. 387-395

/18/ Pöhlandt,K.: Entwicklung eines vereinheitlichten Stauchversuches zur Aufnahme von Fließkurven, Seminar Neuere Entwicklungen in der Massivumformung, Stuttgart 7-8/6/1983

/19/ Nester,W., Pöhlandt,K.: Ermittlung von Fließkurven in verschiedenen Ausführungsformen des Stauchversuches, Rheol. Acta 21 (1982), S. 4o9-412

/2o/   Herbertz,R., Wiegels, H.: Der Unterschied zwischen Zug-
        und Druckfließkurven, gedeutet durch den hydrostatischen
        Druckeinfluß, Arch. Eisenhüttenwes. 51 (198o), S. 413-
        416

/21/   Lach, E., Pöhlandt,K.: Prüfung des plastischen Verhal-
        tens metallischer Werkstoffe im Torsionsversuch, Draht
        33 (1982), S. 689-693

/22/   Pöhlandt,K., Tekkaya, A.E., Lach,E.: Prüfung des plasti-
        schen Verhaltens metallischer Werkstoffe in Torsions-
        versuchen, Z. Werkstofftechnik 14 (1983), S. 181-189

/23/   Gologranc,F.: Beitrag zur Ermittlung von Fließkurven im
        kontinuierlichen hydraulischen Tiefungsversuch, Bericht
        aus dem Institut für Umformtechnik Stuttgart, Nr. 31,
        Verlag Girardet, Essen 1975

/24/   Schott,K.: Plastische Verformung und Stabilität beim
        hydraulischen Tiefungsversuch, Diss. TU Braunschweig
        1973

/25/   Sawada,T.: Error estimates during stress-strain diagram
        measured with hydraulic bulge test, Dep of Mech Engng,
        Tokyo Univ of Agriculture and Technology, Priv Rep 1981

/26/   Marciniak,Z., Kolodziejski,J.: Assessment of sheet metal
        failure sensitivity by method of torsioning the rings,
        J Proc 7th Biannual Congr of the IDDRG, Amsterdam 1972,
        S. 61-64

/27/   Pöhlandt,K., Tekkaya, A.E.: Aufnahme der Fließkurven
        dünner Bleche im ebenen Torsionsversuch, Arch. Eisen-
        hüttenwes. 53 (1982), S. 421-426

/28/   Dripke,M., Wörner,H.P.: Senkrechte Anisotropie r und
        Verfestigungsexponent n, Bänder Bleche Rohre 2o (1979),
        S. 286-291

/29/   Michaelis, E.E.: The r and n test for every day testing
        in the sheet metal industry, Sheet Metal Ind. 56 (1979),
        S. 936-941

/3o/   Bauer,D.: Grundversuche der Metallumformung, Metall 32
        (1978), S. 776-781

/31/   Panknin,W., Bach,M.: Anleitung zur experimentellen Be-
        stimmung der Fließkurven metallischer Werkstoffe, DFBO-
        Mitt. 1972, S. 15-22

/32/   Pawelski,O.: Vergleichende Wertung der Prüfverfahren
        für die Warmumformbarkeit von Metallen, in: Warmumfor-
        mung und Warmfestigkeit, Symp. Bad Nauheim 1975, Dt.
        Ges.f. Metallkunde (DGM), Oberursel 1976

/33/   VDI-Richtlinie 32oo: Fließkurven metallischer Werkstof-
        fe, Grundlagen, Oktober 1978

/34/   Krause, U.: Vergleich verschiedener Verfahren zum Be-
        stimmen der Formänderungsfestigkeit bei Raumtemperatur,
        Stahl und Eisen 83 (1963), S. 1626-164o

/35/   (Hrsg.)Lange,K.:Lehrbuch der Umformtechnik Bd.1:Grundla-
        gen,1.Aufl.,Springer-Verlag, Berlin/Heidelb./N.Y.1972

/36/   Rao,K.P., et al.: Flow curves and deformation of mate-
        rials at different temperatures and strain rates, J.
        Mech.Working Technol. 6 (1982), S. 63-88

/37/   Tekkaya,A.E., Pöhlandt,K., Dannenmann,E.: Methoden zur
        Bestimmung der Fließkurven von Blechwerkstoffen, Blech
        Rohre Profile 29 (1982), S. 354-36o

/38/   Schmidt,W.:Die senkrechte Anisotropie von Feinblechen,
        Blech Rohre Profile 25 (1978), S. 271-275

/39/ Reissner,J.: Fließkurven und Fließortkurven,in:(Hrsg.)
Lange,K.: Lehrbuch der Umformtechnik, Bd. 1:Grundlagen,
2.Aufl., Springer-Verlag, Berlin/Heidelberg/New York,
demnächst

/4o/ Tozawa,Y.: Recommended method in Japan for testing cold
upsettability, ICFG-Vollversammlung, Stuttgart 1979

/41/ Dannenmann,E., Blaich,M.: Verfahren zur Prüfung der
Kaltstauchbarkeit , Draht 28 (1978), S. 7o2-7o6

/42/ Weber,K.-H.: Der Warmtorsionsversuch und seine Aussage
als Maß für das Umformverhalten der Stähle bei höherer
Temperatur, Habil.-schrift, BA Freiberg 1968

/43/ DIN 5o1o1, Blatt 1: Tiefungsversuch an Blechen und Bändern mit einer Breite von $\geq$ 9omm (nach Erichsen),
Dickenbereich o,2mm bis 2mm, August 1961 ; Blatt 2:
Tiefungsversuch an Blechen und Bändern mit einer Breite
von $\geq$ 9omm (nach Erichsen), Dickenbereich über 2mm bis
3mm, August 1961

/44/ DIN 5o1o2: Tiefungsversuch an schmalen Bändern (nach
Erichsen), Breitenbereich 3omm bis unter 9omm, Januar
1963

/45/ Sonne,H.-M., Müschenborn, W.: Kerbzugversuch und Streck-
biegeversuch- Prüfverfahren zur Kennzeichnung der Kalt-
umformbarkeit warm- und kaltgewalzter Flachprodukte,
Arch. Eisenhüttenwes. 5o (1979), S. 5o3-5o8

/46/ Pöhlandt,K.: A note on the notched tensile elongation
and anisotropic ductility of sheet metal, Materialprüf.
24 (1982), S. 83-86

/47/ Schaub,W.: Untersuchungen über das Versagen von Blechen
beim scharfkantigen 18o°-Biegen, Bänder Bleche Rohre 2o
(1979), S. 458-461

/48/   Keeler,S.P.: Determination of forming limits in automo-
         tive stampings, Society of Automotive Engineers (1965),
         Nr. 65o535, S. 1-9

/49/   Goodwin, G.M.: Application of strain analysis to sheet
         metal forming problems in the press shop, Society of
         Automotive Engineers (1968), Nr. 68oo93, S. 38o-387

/5o/   Nakazima, K., et al.: Study on the formability of steel
         sheet, Ytawa Technical Report Nr. 264 (1968), S. 8517-
         853o

/51/   Hasek, V.: Anwendung von Grenzformänderungs-Schaubildern,
         Ind.-Anz. 99 (1977), S. 343-347

/52/   Hasek,V.: Untersuchung und theoretische Beschreibung
         wichtiger Einflußgrößen auf das Grenzformänderungsschau-
         bild, Blech Rohre Profile 25 (1978), S. 213-22o, 281-
         292, 493-499, 619-627

/53/   DIN 476o: Begriffe für die Gestalt von Oberflächen,
         Juli 196o

/54/   DIN 4762: Oberflächenrauheit, Entwurf, Mai 1978

/55/   DIN 4768: Ermittlung der Rauheitsmeßgrößen R_a,R_z,R_{max}
         mit elektrischen Tastschnittgeräten, August 1971

/56/   DIN 4771: Messung der Profiltiefe P_t von Oberflächen,
         April 1977

/57/   DIN 4772:Elektrische Tastschnittgeräte zur Messung der
         Oberflächenrauheit nach dem Tastschnittverfahren,
         November 1979

/58/   Peters,J., et al.: Assessment of surface topology analy-
         sis techniques, Annals of the CIRP 28 (1979), S. 539-554

/59/   Bodschwinna,H.: Auswirkung der Tastspitzengeometrie auf
        die industrielle Rauheitsmessung, Techn. Messen 47 (198o)
        S. 21-28

/6o/   Sayles,R.S., Thomas,T.R.: Measurements of the statistical
        microgeometry of engng. surfaces, Trans ASME, J Lubrica-
        tion Eng 1o1 (1979), S. 4o9-418

/61/   (Hrsg.)Blumenauer, H.: Werkstoffprüfung, Verlag für
        Grundstoffindustrie, Leipzig 1977

/62/   Storbeck, F., Edelmann, Chr.: Oberflächenanalysenverfah-
        ren, Die Technik 33 (1978), S. 226-23o

/63/   Kollek,H.: Verfahren zur schnellen Erkennung von Verun-
        reinigungen auf Metalloberflächen, Metall 33 (1979), S.
        247-25o

/64/   BDS-Fachbuch Blech, Bd.1, 7.Aufl., Bundesverband Deutsch.
        Stahlhandel 1978

/65/   Schorsch,H.: Gütebestimmung an technischen Oberflächen,
        Wiss. Verlagsges., Stuttgart 1971

/66/   Krämer,H.: Oberflächenmessung mit einem Abdruckverfahren,
        Ind.-Anz. 1oo (1978), Nr. 1o2, S. 3o-31

/67/   Droscha,H.: Vollautomatische Oberflächenprüfung von
        Flachmaterial mit Laserstrahl, Metall 31 (1977), S.1o84

/68/   DIN 5o1o3, Blatt 1: Prüfung metallischer Werkstoffe,
        Härteprüfung nach Rockwell, Verfahren C,A,B,F, Dezember
        1972, Blatt 2: Härteprüfung nach Rockwell, Verfahren N
        und T, Oktober 1973

/69/   DIN 5o1o3, Blatt 3: Härteprüfung nach Vickers, Prüfkraft-
        bereich 49 bis 98o N (5 bis 1oo kp), Dezember 1972,
        Blatt 4: Härteprüfung nach Vickers, Prüfkraftbereich

1,96 bis 49 N (o,2 bis 5 kp) (Kleinlastbereich), Dezember 1972

/7o/ DIN 5o351: Härteprüfung nach Brinell, Januar 1973

/71/ DIN 5o115: Kerbschlagbiegeversuch , Februar 1975

/72/ ASTM Standard E 399-81: Plane-strain fracture toughness of metallic materials, American Society of Testing and Materials, Philadelphia Pa., USA 1981

/73/ DIN 5o1oo: Dauerschwingversuche, Begriffe, Zeichen , Durchführung, Auswertung, Januar 1953

/74/ DIN 5o113: Umlaufbiegeversuch, Dezember 1952

/75/ DIN 5o118: Zeitstandversuch, Dezember 1952

/76/ DIN 5o119: Standversuch, Begriffe, Zeichen, Durchführung, Auswertung, Dezember 1952

/77/ DIN 5o9o5, Blatt 1. Chemische Korrosionsuntersuchungen, Allgemeines, Januar 1975, Blatt 2: Chemische Korrosionsuntersuchungen, Korrosionsgröß en bei gleichmäßiger Flächenkorrosion, Blatt 3: Chemische Korrosionsuntersuchungen, Korrosionsgrößen bei ungleichmäßiger Korrosion ohne zusätzliche mechanische Beanspruchung, alle Januar 1975

/78/ DIN 5o9o8: Prüfung von Leichtmetallen: Spannungskorrosionsversuche, Juli 1974

/79/ DIN 5o911 : Prüfung von Kupferlegierungen. Quecksilbernitratversuch, Juni 1974

/8o/ DIN 5o914: Prüfung nichtrostender Stähle auf Beständigkeit gegen interkristalline Korrosion. Kupfersulfat-Schwefelsäure-Verfahren, September 197o

/81/ DIN 5o915: Prüfung von unlegierten und niedriglegierten
      Stählen auf Beständigkeit gegen interkristalline Span-
      nungsrißkorrosion, Juli 1975

/82/ DIN 5oo21: Sprühnebelprüfungen mit verschiedenen Natrium-
      chloridlösungen, Mai 1975

/83/ ASTM Standard G 36-73: Performing stress-corrosion crak-
      king tests in a boiling magnesium chloride solution,
      American Society of Testing and Materials, Philadelphia
      Pa., USA 1973

/84/ ASTM Standard E 112-82: Standard methods for determining
      average grain size, American Society of Testing and
      Materials, Philadelphia Pa., USA 1982

/85/ DIN 5o6oo: Metallographische Gefügebilder. Abbildungs-
      maßstäbe und Formate, Dezember 1952

/86/ Schumann, H.: Metallographie, 9. Aufl., Verlag für
      Grundstoffindustrie, Leipzig 1975

/87/ Rostoker,W., Dvorak, J.R.: Interpretation of Metallogra-
      phic Structures, 2nd ed., Academic Press, New York 1977

/88/ (Hrsg.) Kommission der Europ. Gemeinschaft: De Ferri
      Metallographia, 3Bde, Verlag Stahleisen, Düsseldorf 1966

/89/ Stahl-Eisen-Prüfblätter des VdEh, Blatt 152o 3.78:Mik-
      roskopische Prüfung der Carbidausbildung in Stählen und
      Bildreihen, Düsseldorf 1978

/9o/ Stahl-Eisen-Prüfblätter des VdEh, Blatt 157o 8.71: Mik-
      roskopische Prüfung von Stählen auf nichtmetallische
      Einschlüsse mit Bildreihen, Düsseldorf 1971

/91/ Schippers,M.: Metallographie von Aluminium-Legierungen,
      in: Lehrgang Metallographische Untersuchungsmethoden III,

Technische Akademie Esslingen (TAE), 25-27/1o/1978

/92/   Beckert,M., Klemm, H.: Handbuch der metallographischen
       Ätzverfahren, Verlag für Grundstoffindustrie, Leipzig
       1962

/93/   Häßner,F.: Verfahren zur Texturbestimmung, in(Hrsg.)
       Grewen,J., Wassermann,G.: Texturen in Forschung und Pra-
       xis, Symp. Clausthal-Zellerfeld, 2-5/1o/1968, Springer-
       Verlag, Berlin/Heidelberg/New York 1969

/94/   Bunge,H.-J.: Methods of ODF Calculation, in (Ed.) Gott-
       stein, G., Lücke, K.:Textures of Materials, Proc. Symp.
       Aachen, 28-31/3/1978, Springer-Verlag, Berlin/Heidel-
       berg/ New York 1978

/95/   Stüwe,H.-P., Vibrans, G.: Feinstrukturuntersuchungen in
       der Werkstoffkunde, Bibliographisches Institut, Mann-
       heim/Wien/Zürich 1974

/96/   V. Heimendahl, M.: Einführung in die Elektronenmikrosko-
       pie, Vieweg- Verlag, Braunschweig 197o

/97/   Leykamm, H.: Berechnung der Fließkurve vob Fließpreß-
       stählen und der Festigkeit, Bruchdehnung und Bruchein-
       schnürung kaltumgeformter Werkstücke aus der Werkstoff-
       analyse nach Wagenbach, Draht 29 (1978), S. 648-651

/98/   Jonck,R., et al.: Einfluß der Legierungselemente auf das
       Kaltumformverhalten, Zeitschrift für wirtsch. Fertigung
       (ZwF) 69 (1974), S. 419-424

/99/   VDI-Arbeitsblatt 32oo bis 32o2: Fließkurven metallischer
       Werkstoffe, Oktober 1978

/1oo/ (Hrsg.) Deutsche Gesellschaft für Metallkunde (DGM):
       Atlas der Warmformgebungseigenschaften von Nochteisen-
       metallen, Bd.1: Aluminiumwerkstoffe, Bd.2: Kupferwerk-

stoffe, DGM, Oberursel 1978

/1o1/ Neuberger,F., et al.: Klassifikation gebräuchlicher
Schmiedewerkstoffe durch Stauchversuche, Masch.-B.-
Techn. 7 (1958), S. 249-254

/1o2/ Hensel,A., Spittel,Th.: Kraft- und Arbeitsbedarf bild-
samer Formgebungsverfahren, Verlag für Grundstoffindustrie, Leipzig 1976

/1o3/ (Hrsg.) Verein Deutscher Eisenhüttenleute: VdEh-Werk-
stoffdaten, Verlag Stahleisen, Düsseldorf

/1o4/ Stahlschlüssel, 13. Auflage , Verlag Stahlschlüssel
Wegst KG, Marbach 1983

/1o5/ Wellinger,K., Gimmel, P., Bodenstein, M.: Werkstoff-
Tabellen der Metalle, 7. Aufl., Kröner Verlag, Stuttgart 1972

/1o6/ Tichy,J., Gautschi,G.: Piezoelektrische Meßtechnik,
Springer- Verlag, Berlin/Heidelberg/New York 198o

/1o7/ Hoang-Vu,Kh.: Möglichkeiten und Grenzen des Kaltgesenkschmiedens als eine fertigungstechnische Alternative
für kleine, genaue Formteile, Bericht aus dem Institut
für Umformtechnik Stuttgart, Nr. 65, Springer-Verlag,
Berlin/Heidelberg/New York 1982

/1o8/ Siebel,E., Lueg,W.: Untersuchungen über die Spannungsverteilung im Walzspalt, Mitt. KWI Eisenforsch. 15
(1933), S. 1-14

/1o9/ Dohmann,F.: Die Messung der mechanischen Kontaktnormalspannung in der Wirkfuge Werkzeug-Werkstück bei Umformverfahren, Bericht aus dem Institut für Umformtechnik
Stuttgart, Nr. 27, Verlag Girardet, Essen 1974

/11o/   Matsubara,S., Kudo,H.: Determination of pressure dis-
        tribution over tool surface in cold forging with a
        simple sensor, Annals of the CIRP 25 (1977), S.95-1oo

/111/   Steck, E., Geiger,M.: Plastizitätstheoretische Grundla-
        gen, in (Hrsg.) Lange, K.: Lehrbuch der Umformtechnik,
        Bd. 1:Grundlagen, 1. Aufl. , Springer-Verlag , Berlin/
        Heidelberg /New York 1972

/112/   Leopold, J.: Bestimmung kontinuumsmechanischer und
        thermodynamischer Kenngrößen in der Umformzone mittels
        Visioplastizität, Neue Hütte 28 (1983), S. 26-29

/113/   Keeler,S.P.: Circular grid system - a valuable aid for
        evaluating sheet metal formability, Sheet Metal Ind.
        45 (1968), S. 633-641

/114/   Schelosky,H.: Verfahren zur Prüfung der Umformeignung
        von Blechen, in (Hrsg.) Lange,K.:Lehrbuch der Umform-
        technik , Bd.3: Blechumformung, 1. Aufl., Springer-
        Verlag, Berlin/Heidelberg/New York 1975

/115/   DIN 6583: Begriffe der Zerspanbarkeit. Standbegriffe,
        Entwurf, Mai 1979

/116/   Stahl-Eisen-Prüfblätter des VdEh, Blatt 1168-69: Zer-
        spanversuche -Zerspankraftmessung im Drehversuch, 2.
        Ausgabe, Düsseldorf 1969

/117/   Stahl-Eisen-Prüfblätter des VdEh, Blatt 1178-69: Zer-
        spanversuche - Spanbeurteilung, 2. Ausgabe, Düsseldorf
        1969

/118/   Nehl,E.: Untersuchungen zum Halbwarmfließpressen von
        Automatenstählen, Bericht aus dem Institut für Umform-
        technik Stuttgart, Nr. 7o, Springer-Verlag, Berlin/
        Heidelberg/New York /Tokyo 1983

/119/   Rastegaev,M.V.: Neue Methode der homogenen Stauchung
         von Proben zur Bestimmung der Fließspannung und des
         Koeffizienten der inneren Reibung (russ.), Zavod.Lab.
         (1940), S. 354

/12o/   Krokha, V.A.: A method of determining the flow stress
         in compression to high degrees of plastic deformation,
         Ind. Lab. 4o (1974), S. 754-758

/121/   Pöhlandt,K.: Stauchversuch zur Ermittlung von Fließkur-
         ven nach Rastegaev, Ind.-Anz. 1o1 (1979), 48, S.28-29
         (HGF 79/26)

/122/   Wilhelm,H.: Untersuchungen über den Zusammenhang zwi-
         schen Vickershärte und Vergleichsformänderung bei Kalt-
         umformvorgängen, Bericht aus dem Institut für Umform-
         technik Stuttgart, Nr.9, Verlag Girardet, Essen 1969

/123/   Zeller,R.: Änderung der Werkstoffeigenschaften beim
         Ziehen von zylindrischen Hohlkörpern aus austenitischen
         und ferritischen nichtrostenden Stählen, Bericht aus
         dem Institut für Umformtechnik Stuttgart , Nr. 42, Ver-
         lag Girardet, Essen 1976

/124/   Kerspe,J.H.: Abstreckgleitziehen von nichtrostenden
         austenitischen Stählen, Bericht aus dem Institut für
         Umfortechnik Stuttgart, Nr. 53, Springer-Verlag, Ber-
         lin/Heidelberg/New York 198o

/125/   Macherauch,E.: Praktikum in Werkstoffkunde, 2. Aufl.,
         Vieweg- Verlag, Braunschweig 1972

/126/   Bunge,H.-J.: Mathematische Methoden der Texturanalyse,
         Akademie-Verlag, Berlin 1969

/127/   Kloos,K.H.: Eigenspannungen, Definition und Entstehungs-
         ursachen, Z. Werkstofftechnik 1o (1979), S. 293-3o2

/128/ (Hrsg.) Blumenauer,H.: Werkstoffprüfung, Verlag für Grundstoffindustrie, Leipzig 1977

/129/ Dorfschmidt,E.: Eigenspannungsuntersuchungen beim Voll-Vorwärts-Fließpressen von Stahl, Diss. Univ. Hannover 1982

/13o/ Spannungsermittlung mit Röntgenstrahlen, Sonderheft, Härt.-Techn. Mitt. 31 (1976)

/131/ Bölle, H.,Hauk,V.: Röntgenographische Ermittlung von Eigenspannungen in texturierten Werkstoffen, Z. Metallkde 7o (1979), S. 682-689

/132/ Lee,E.H., et al.: Stress and Deformation Analysis of Metal-Forming Processes, in: (Ed.) Lippmann, H.: Metal Forming Plasticity, Springer-Verlag, Berlin/Heidelberg/ New York 1979

/133/ Kudo,H., Matsubara,S.: Joint Examination Project of Validity of Various Numerical Methods for the Analysis of Metal Forming Processes, in (Ed.) Lippmann,H.: Metal Forming Plasticity, Springer- Verlag, Berlin/Heidelberg/New York 1979

/134/ Tekkaya, A.E.:Beitrag zur Berechnung von Umformeigenspannungen in Fließpreßteilen, in: Seminar Neuere Entwicklungen in der Massivumformung, Stuttgart, 7-8/6/ 1983

/135/ Pawelski,O.: Beitrag zur Ähnlichkeitstheorie in der Umformtechnik, Arch. Eisenhüttenwes. 35 (1964), S. 27-36

/136/ Kast,D.: Modellgesetzmäßigkeiten beim Rückwärtsfließpressen geometrisch ähnlicher Näpfe, Bericht aus dem Institut für Umformtechnik, Nr. 13, Verlag Girardet, Essen 1969

/137/   Berns, H.: Metallurgisch bedingter Größeneinfluß in
        Stählen, Ind.-Anz. 1o2 (198o), 93, S. 28-31

/138/   Leykamm,H.: Beitrag zur Arbeitsgenauigkeit des Kalt-
        massivumformens, Bericht aus de Institut für Umform-
        technik Stuttgart, Nr. 57, Springer-Verlag, Berlin/
        Heidelberg/New York 198o

/139/   Schlicht, H.: Einfluß der Stahlherstellung auf das Er-
        müdungsverhalten von Bauteilen bei kräftefreier und
        kräftegebundener Oberfläche, Zeitschrift für wirtsch.
        Fertigung (ZwF) 73 (1978), S. 589-598

/14o/   Munz, D., Schwalbe, K., Mayr,P.: Dauerschwingverhalten
        metallischer Werkstoffe, Vieweg- Verlag, Braunschweig
        1971

/141/   (Hrsg.) Schott, G.: Werkstoffermüdung, Verlag für
        Grundstoffindustrie Leipzig 1977

/142/   Maier, H.-J.: Über den Einfluß einer Ka ltverformung auf
        die Zeitfestigkeit biegebeanspruchter glatter Proben,
        Techn.-wiss. Ber. MPA Stuttgart (1975), Heft 75-o2

/143/   Strigens, P.: Einfluß der Oberflächenverfestigung auf d
        die Dauerfestigkeit von Stählen, VDI-Z. 114 (1972),
        S. 1193-1199

/144/   Jonck,R.: Oberflächen- und Randschichteinfluß auf die
        Festigkeitseigenschaften massivgeformter Werkstücke,
        Zeitschrift für wirtsch. Fertigung (ZwF) 16 (1981),
        S.496-5o2

/145/   Tauscher,H.: Berechnung der Dauerfestigkeit, Einfluß
        von Werkstoff und Gestalt, 6. Aufl. Fachbuchverlag
        Leipzig 196o

/146/   Flemming,G.: Mechanische Eigenschaften von Stahl bei
         statischer und wechselnder Beanspruchung nach einer
         Massivumformung, Diss. TH Darmstadt 1972

/147/   Feldmann,H., Hagedorn,K.E.: Grundlagen der Zeitfestig-
         keit, Stahl und Eisen 97 (1977), S.1o99-11o6

/148/   Knolle,B.: Schwingfestigkeit von Schmiedeteilen, Ind.-
         Anz. 1o1 (1979), 18, S. 41-42 (HGF 79/8)

/149/   Lehr,E.: Die Abkürzungsverfahren zur Ermittlung der
         Schwingfestigkeit von Materialien, Diss TH Stuttgart
         1925

/15o/   Harig, H., Frank,E.: Temperaturmessung als einfache
         Methode zur Charakterisierung des Ermüdungsverhaltens
         von Stählen, Zeitschrift für wirtsch. Fertigung (ZwF)
         76 (1981), S. 185-188

Turbinenscheiben für Flugtriebwerke: Beanspruchung, Werkstoffe, Fertigungsverfahren

H. Wilhelm, MTU München

Summary:

Aero engines have to fulfill very strong requirements concerning safety and reliability. This applies especially to rotating components such as compressor and turbine discs. Stressing conditions for discs are explained. Stress maximum is at the hub of the disc. This region as well as notches, holes etc. are susceptible to low-cycle fatigue.

The most important titanium alloys for compressor discs are described. Their mechanical and forging properties are explained.

Due to the high operation temperature turbine discs must be made of nickel-base alloys (superalloys). The reasons for their high-temperature strength are explained.

Discs made of "conventional" superalloys (Inconel 718, Waspaloy) are usually produced by conventional forging applying hydraulic presses.

Discs of advanced high-strength superalloys (AP1, René 95 etc.) are exclusively made of P/M-preforms. Possible processing sequences for production of P/M-discs are:

- HIP plus conventional forging
- HIP plus hot forging
- HIP plus isothermal forging
- HIP (without subsequent forging).

The technical and economical aspects of the different production routes are discussed by means of selected examples.

1 Einleitung

Luftfahrtkonstruktionen sind gekennzeichnet durch

- einen extremen Leichtbau, in dem Werkstoffe mit hoher
  gewichtsspezifischer Festigkeit eingesetzt und bis an
  die zulässige Grenze ihres Festigkeitspotentials ausgenützt werden.
- eine hohe erforderliche (und auch erreichte) Betriebs-
  sicherheit der Bauteile.

Dies erfordert die sichere Reproduzierbarkeit der Werkstoffeigenschaften sowie aller Fertigungs- und Kontroll-
schritte.
Darüber hinaus unterliegen zahlreiche Triebwerksteile noch
einer hohen thermischen und korrosiven Belastung durch
Heißgas. Bei modernen Triebwerken beträgt die Gastempera-
tur am Brennkammeraustritt über 1400 °C. Die Temperatur
der vom Heißgas beaufschlagten Bauteile, vor allem der
Turbinenschaufeln, muß deshalb durch Kühlsysteme so weit
abgesenkt werden, daß der Werkstoff noch eine ausreichende
Festigkeit besitzt.
Zur Zeit sind bei der Entwicklung neuer Triebwerke starke
Tendenzen zur Verminderung des Kraftstoffverbrauches, zur
Verminderung der Lärm- und Schadstoffemission und zur
kostengünstigeren Fertigung erkennbar.
Die Verminderung des Kraftstoffverbrauches wird dabei
durch eine weitere Erhöhung der Gastemperatur und des
Druckverhältnisses (Drehzahl) erreicht, die zu einer noch
höheren mechanischen und thermischen Belastung der Bau-
teile führt.

2 Beanspruchungsverhältnisse von Verdichter- und Turbi-
nenscheiben

Verdichter- und Turbinenscheiben stellen außer der Be-
schaufelung die höchst beanspruchten Triebwerksteile dar.

Sie werden durch hohe Fliehkräfte bei örtlichen Werk-
stofftemperaturen bis ca. 600 °C (Turbine) belastet und
haben außerdem noch Temperatur-Gradienten zu ertragen.

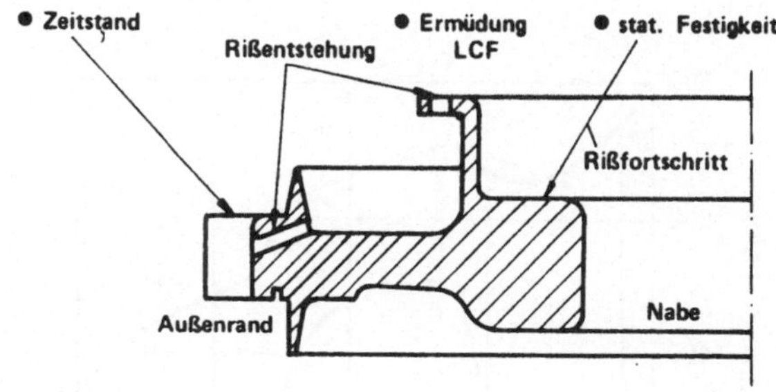

Bild 1: Betriebsbeanspruchungen an einer
Turbinenscheibe.

Anhand der in Bild 1 gezeigten Prinzipskizze sollen die
Betriebsbeanspruchungen einer Turbinenscheibe erläutert
werden:
Der Innenrand (Lochleibung) hat die höchsten Spannungen zu
ertragen. Sie können bei modernen Triebwerken ca. 1000 MPa
betragen. Deshalb ist hier eine hohe statische Festigkeit
erforderlich. Am Außenrand der Scheibe ist die Nennspannung niedriger, dafür aber die Temperatur höher (max. ca.
600 °C). Vom Werkstoff werden hier deshalb eine hohe
Warmfestigkeit und günstige Zeitstandeigenschaften gefordert.

Alle schroffen Änderungen des Querschnitts bzw. der Bau-
teilkontur, z. B. Kanäle zur Zuführung von Kühlluft in die
Tannenbaum-Nuten, oder Bohrungen für Befestigungselemente,
erzeugen örtliche Spannungsspitzen und sind deshalb besonders empfindlich gegen LCF-Risse. Eine sehr sorgfältige
mechanische Bearbeitung dieser Bereiche zur Erzeugung
einer glatten, grat-, riß- und riefenfreien Oberfläche ist
an diesen Stellen besonders wichtig. Bisweilen wird an

manchen gefährdeten Querschnitten kugelgestrahlt, um
Druckeigenspannungen in die Oberfläche einzubringen.
Besonders LCF-gefährdet ist auch der Innenrand der Scheibe, der die höchste Nennspannung des ganzen Bauteils zu
ertragen hat.

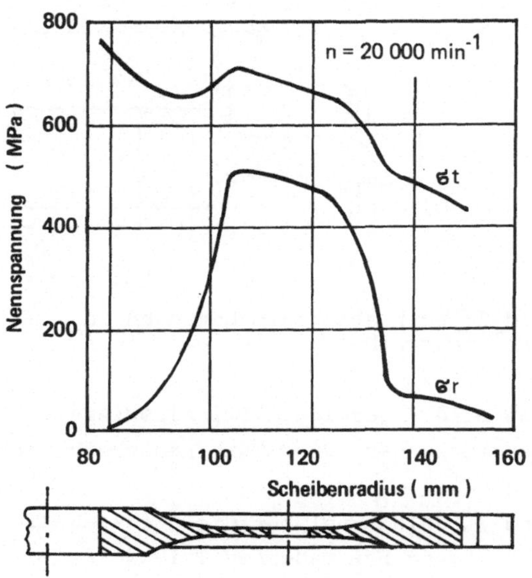

Bild 2: Radial- und Tangentialspannung
an einer Testscheibe aus Inconel 718.

Bild 2 zeigt den Verlauf der Radial- und Tangentialspannung an einer Schleuderscheibe ohne Berücksichtigung
von Kerbspannungen, die örtlich die Streckgrenze überschreiten können.
Turbinenscheiben von modernen Triebwerken werden im allgemeinen aus Gewichtsgründen nur auf Zeitfestigkeit, nicht
aber auf Dauerfestigkeit ausgelegt. Die typischen Lebens-
dauern betragen $0,5 \cdot 10^4$ bis $3,0 \cdot 10^4$ Lastwechsel, wobei
die höhere Zahl für langlebige zivile Triebwerke gilt.

Unter einem Lastwechsel kann vereinfachend die Zeit vom
Starten des Triebwerks über Start, Reiseflug und Landung
bis zum Abschalten des Triebwerks verstanden werden.

Bei Triebwerken der nächsten Generation werden die Scheiben aufgrund der vorgesehenen Drehzahl- und Temperatur-
steigerungen nochmals beträchtlich höher belastet werden,
als dies bei schon verfügbaren Triebwerken der Fall ist.
Dies erfordert zwingend den Einsatz noch leistungsfähige-
rer Werkstoffe, von denen heute bereits einige an der
Schwelle zur großtechnischen Produktion stehen.

3 Werkstoffe für Verdichter- und Turbinenscheiben

3.1 Werkstoffauswahl, wichtige Legierungen

Die wichtigsten Werkstoffgruppen für Scheiben sind Titan-
und Nickelbasis-Legierungen (Superlegierungen). Erstere
kommen aufgrund ihrer begrenzten Einsatztemperatur nur für
den Verdichter in Betracht.
In den letzten Verdichterstufen moderner Triebwerke können, bedingt durch die adiabate Verdichtung, Gastempe-
raturen von über 600 °C auftreten. Deshalb kommen hierfür
sowie für sämtliche Scheiben in der Turbine nur Nickelbasislegierungen zur Anwendung.

Titanlegierungen:
Titanlegierungen werden entsprechend den nach der Verarbeitung im Gefüge vorherrschenden Phasen in α-, $(\alpha+\beta)$-
und β-Legierungen unterteilt. α-Legierungen weisen ein
hexagonales, β-Legierungen ein kubisch raumzentriertes
Gitter auf.
Die für Verdichterscheiben wichtigen Legierungen gehören
ausnahmslos zu den $(\alpha+\beta)$-Legierungen.
Die wichtigsten Legierungen für Verdichterscheiben sind

- Ti Al6 V4
- Ti Al6 Sn2 Zr4 Mo2
- Ti Al6 Mo 0,5 Zr5 Si 0,2 (= Ti 685).

Legierung	20 °C			400 °C			500 °C		
	Rp 0,2 MPa	Rm MPa	A5 %	Rp 0,2 MPa	Rm MPa	A5 %	Rp 0,2 MPa	Rm MPa	A5 %
Ti Al 6V4	965	1000	14	560	675	16	465	585	30
Ti Al 6Sn2 Zr4 Mo2	940	1000	15	570	730	15	530	680	18
Ti 685	950	1000	11	600	745	15	540	690	14

Tabelle 1: Festigkeit geschmiedeter
Titanlegierungen.

In Tabelle 1 finden sich einige statische Festigkeitswerte
dieser Legierungen. Bei Raumtemperatur sind die Festigkeitswerte aller drei Werkstoffe ähnlich, bei 500 °C fällt
jedoch die Festigkeit von Ti Al6 V4 gegenüber den anderen
Legierungen ab.

Nickelbasis-Legierungen (Superlegierungen)

Die Superlegierungen sind speziell für den Einsatz bei
höheren Temperaturen ausgelegt. Der entscheidende Beitrag
zur Erhöhung der Hochtemperaturfestigkeit resultiert aus
der Blockade der Versetzungen durch eine kohärente Aus-
scheidungsphase (γ'), die im wesentlichen aus Ni_3 (Al, Ti)
besteht. Durch eine Gitterverspannung mit Hilfe größerer
und schwerer Atome (z. B. W, Mo) kann die Hochtemperaturfestigkeit ebenfalls erhöht werden, doch sind dieser
Möglichkeit bei Scheibenwerkstoffen aus Gewichtsgründen
enge Grenzen gezogen.
Mit zunehmendem γ'-Gehalt werden auch das metallurgische
Verhalten und die Umformbarkeit von Superlegierungen
zusehends "schwieriger".

Angaben in Gewichts — %										
Werkstoff	C	Cr	Ni	Co	Mo	W	Fe	Ti	Al	andere
Inconel 718	0,04	18,5	52,0	—	3,0	—	Rest	1,0	0,6	5,2 Nb + Ta
Waspaloy	0,03	19,5	Rest	13,5	4,5	—	2,0	3,0	1,3	
Astroloy	0,04	15,0	Rest	17,0	5,2	—	0,5	3,5	4,0	
René 95	0,07	13,0	Rest	8,0	3,5	3,50	0,5	2,5	3,5	3,5 Nb + Ta
Inconel 100	0,18	10,0	Rest	15,0	3,0	—	—	4,7	5,5	1,0 V
AF 115	0,05	10,5	Rest	15,0	2,8	5,8	1,0	4,0	3,8	1,90 Nb 0,95 Hf

Tabelle 2: Chemische Zusammensetzung einiger
Superlegierungen für Turbinenscheiben

Bei schmelzmetallurgischer Herstellung von Superlegierungen mit hohem γ'-Anteil bildet sich ein dendritisches,
grobkörniges, inhomogenes und mit Seigerungen behaftetes
Gefüge aus, das nicht oder nur begrenzt umformbar ist. Von
den in Tabelle 2 aufgelisteten Superlegierungen werden nur
Waspaloy und Inconel 718 auf konventionelle Weise, d. h.
durch Abgießen und nachfolgendes Verschmieden von im
Vakuum erschmolzenen Blöcken hergestellt.
"Modernere" Legierungen wie AP1, René 95 oder AF 115
lassen sich nur auf pulvermetallurgischem Wege erzeugen.

Bild 3 zeigt die Warmstreckgrenze einiger wichtiger Le-
gierungen für Turbinenscheiben in Abhängigkeit von der
Temperatur. Besonders bemerkenswert sind hierbei das hohe
Niveau der Kurve für René 95 PM. Bei AF 115 setzt der
Steilabfall der Festigkeit erst oberhalb 700 °C ein. Diese
Legierung ist oberhalb von 600 °C allen zur Zeit verfüg-
baren Scheiben-Legierungen überlegen.
Die dynamischen sowie die bruchmechanischen Eigenschaften
der Legierungen sind für ihre Gebrauchsfähigkeit minde-

stens ebenso wichtig wie die statischen Festigkeitseigenschaften. Diese Daten werden jedoch meist als vertraulich
behandelt und nicht publiziert.

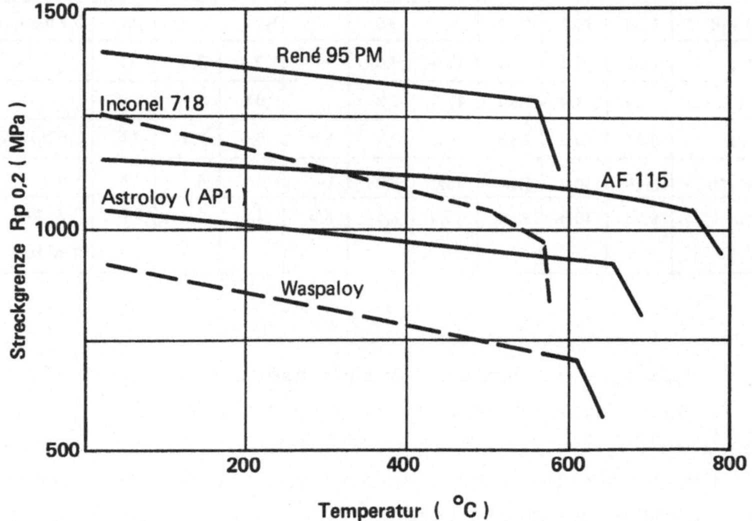

Bild 3: Einfluß der Temperatur auf die Streckgrenze einiger Superlegierungen.

3.2 Einfluß von Werkstoffehlern auf das Ermüdungsver-
halten

Innere Fehlstellen wie Poren, Seigerungen oder nichtme-
tallische Einschlüsse, die im bruchmechanischen Sinn als
Rißkeime anzusehen sind, kommen in allen technischen
Werkstoffen vor. Sie sind durch die Legierungsmetallurgie
und den Herstellprozeß bedingt. Durch zerstörungsfreie
Prüfverfahren lassen sich nicht alle dieser Fehler ermitteln.

Wird ein rißbehaftetes Bauteil einer schwellenden oder
wechselnden Belastung unterworfen, so ergibt sich eine
Rißwachstumskurve [1] gemäß Bild 4.

Man muß dabei drei Bereiche unterscheiden:

a) unterhalb eines Spannungsintensitätsfaktors ΔK_o erfolgt kein Rißwachstum.
b) einen mittleren Bereich, in dem die Rißausbreitungsgeschwindigkeit (bei doppellogarithmischer Darstellung) linear mit ΔK zunimmt. Dieser Bereich läßt sich durch die Beziehung $da/dN = C \cdot \Delta K^n$ beschreiben.
c) einen Instabilitäts-Bereich, in der Nähe von ΔK_c, in dem die Rißwachstumsrate rapid anwächst.

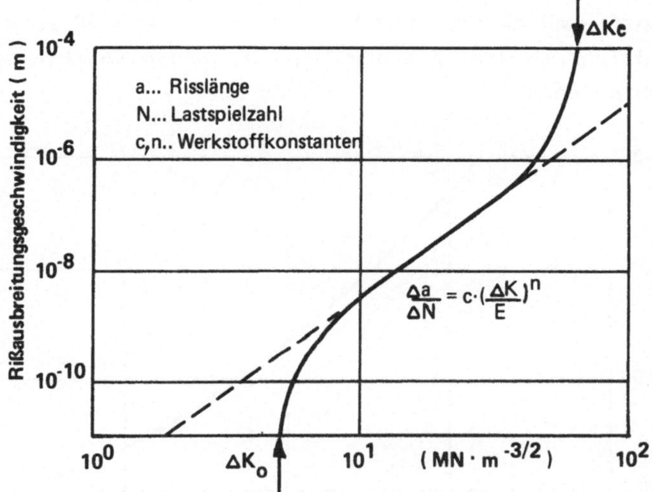

$$\frac{\Delta a}{\Delta N} = c \cdot \left(\frac{\Delta K}{E}\right)^n$$

Bild 4: Ermüdungs-Rißwachstumskurve

Da Turbinenscheiben im allgemeinen nicht auf Dauerfestigkeit ausgelegt werden, ergibt sich für die konstruktive Auslegung die Aufgabe, die Belastung so zu wählen, daß die inneren Risse und Kerben innerhalb der angestrebten Bauteillebensdauer (Lastwechselzahl) die sog. kritische Rißlänge a_c nicht erreichen.
Es wurde verschiedentlich nachgewiesen, daß die Fehlerempfindlichkeit von Scheibenwerkstoffen mit zunehmender

Festigkeit zunimmt [2]. Dies bedeutet, daß sich das Festigkeitspotential der hochfesten Werkstoffe (René 95, AF115) nur dann voll ausnützen läßt, wenn die Fehlergröße reproduzierbar auf niedrige Werte begrenzt werden kann. Andernfalls kann bei hoher Belastung dieser Werkstoffe ein kritisches Rißwachstum mit katastrophalen Folgen eintreten.

In Bild 5 ist für Nickelbasislegierungen der Zusammenhang zwischen angelegter Spannung und zulässiger Fehlergröße zum Erreichen von 10^5 Lastwechseln aufgezeigt. Sind z. B. werkstoff- und herstellungsbedingt Fehler von 50 µm unvermeidlich, so darf die angelegte Spannung 865 MPa nicht überschreiten.

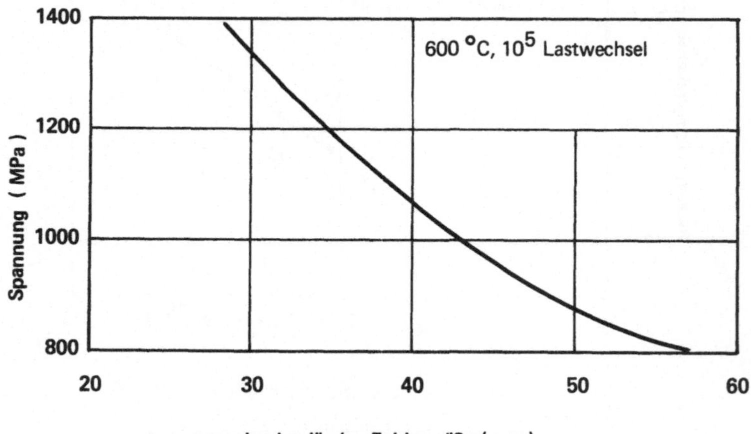

Bild 5: Einfluß der Fehlergröße auf die von Superlegierungen ertragbare Spannung.

Dieselbe Wirkung wie von den genannten inneren Fehlern geht auch von äußeren Kerben und Anrissen am Rand des Bauteils aus. Bei den hochfesten PM-Legierungen kann bereits der Eindruck einer Vickers-Pyramide einen LCF-Riß auslösen. Deshalb ist bei diesen Werkstoffen eine kerb- und kratzerfreie Oberfläche mit geringer Rauhigkeit unabdingbar.

Bild 6 und 7 zeigt für die Knetlegierung Inconel 718 und
die PM-Legierung Udimet 700 je einen typischen inneren
Werkstoffehler. Seigerungen sind in Knetlegierungen mit
hohem γ'-Anteil fast unvermeidlich. Sie können jedoch
durch eine geeignete Schmiedefolge (hoher Umformgrad)
klein gehalten werden (Zerkleinerung der Fehlstellen). PM-
Legierungen sind zwar frei von Seigerungen, durch den
Verdüsungsvorgang können jedoch Abriebpartikel des kera-
mischen Schmelztiegels in den Werkstoff gelangen. Die
einzig sichere Methode zur Vermeidung unzulässig großer
Fremdkörper ist ein entsprechendes Absieben des Pulvers.
Bei modernen PM-Legierungen liegt die Partikelgröße um ca.
50 µm. Bei entsprechender Siebung der Pulver können dann
auch keine größeren Einschlüsse mehr vorhanden sein. Die
Frage, ob bzw. wie stark die LCF-Eigenschaften von PM-
Werkstoffen durch ein Nachschmieden verbessert werden
können, läßt sich nur schwer beantworten. Vor etwa 7 Jahren, als nur verhältnismäßig grobkörnige und stark ver-
unreinigte Pulver zur Verfügung standen, war mit dem
Nachschmieden eine deutliche Verbesserung der Werkstoffeigenschaften verbunden: Große Einschlüsse wurden zer-
kleinert und die mit Oxiden, Karbiden und Verunreinigungen
belegten Partikelgrenzen konnten beseitigt und in ein
weniger grobes Knetgefüge umgewandelt werden.
Bei den heute verwendeten sehr reinen und feinkörnigen
Pulvern liegen die Verhältnisse anders. Der Vorteil des
Nachschmiedens ist dadurch geringer geworden. Jüngere
Untersuchungen deuten darauf hin, daß durch Nachschmieden
die Streubreite der LCF-Festigkeit eingeengt und die
Untergrenze des Streubandes angehoben wird.

Es soll hier nicht verschwiegen werden, daß die Frage, ob
ein Nachschmieden von HIP-Vorformen notwendig und sinnvoll
sei, von verschiedenen Triebwerksherstellern unterschiedlich beurteilt wird und bisweilen ins Philosophische
abschweift. Tatsache ist jedoch, daß die meisten PM-Schei-
ben zur Zeit noch nachgeschmiedet werden. Angesichts einer

auch zukünftig zunehmenden Pulverqualität und -Reinheit
ist davon auszugehen, daß zukünftig der Anteil der durch
einen einstufigen HIP-Prozeß hergestellten Scheiben eher
zu- als abnehmen wird.

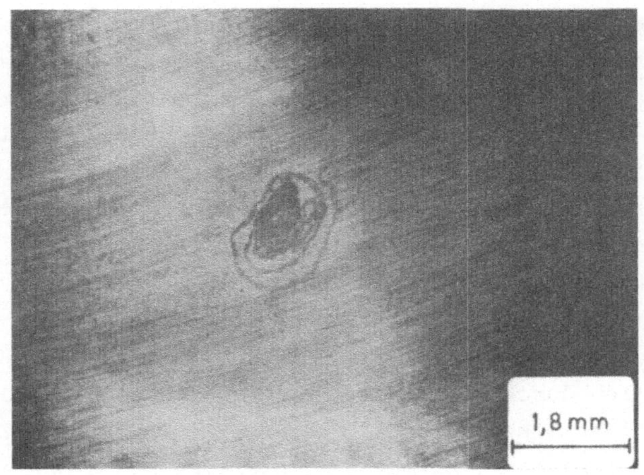

Bild 6: Seigerung in Inconel 718.

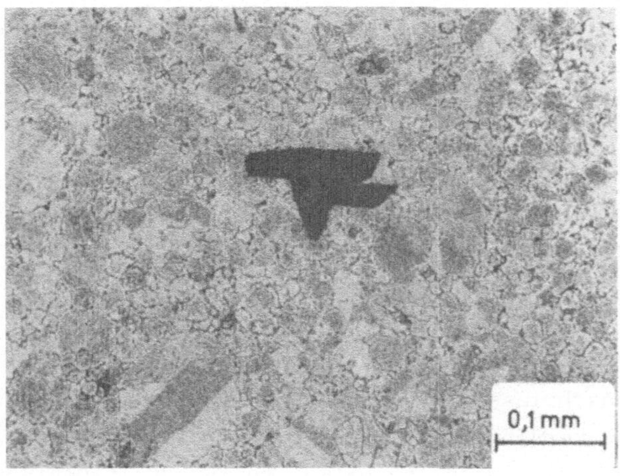

Bild 7: Al_2O_3-Einschluß in Astroloy.

3.3 Umformverhalten von Titanlegierungen

Die meisten Titanlegierungen sind erst oberhalb 850 °C
schmiedbar. Als Schmiederohteile für Turbinenscheiben
werden nur mehrfach im Vakuum umgeschmolzene Gußblöcke
verwendet. Das Vorschmieden der Blöcke erfolgt häufig als
sog. β-Schmieden oberhalb der $(\alpha+\beta)/\beta$-Umwandlungstemperatur, die für die wichtigsten Legierungen zwischen ca.
970 °C und 1010 °C liegt [3]. In diesem Temperaturbereich
weisen die Titanlegierungen eine niedrige Fließspannung
und eine hohe Duktilität auf. Dadurch ergeben sich niedrige Umformkräfte und ein geringer Werkzeugverschleiß.
Beim β-Schmieden stellt sich meist ein martensitisches
nadeliges Gefüge (Widmannstätten) ein.
Ein derartiges Gefüge ist aufgrund der damit verbundenen
ungünstigen Dehnungs- und Einschnürungswerte für Triebwerksteile nicht akzeptabel [4].

Um das β-Gefüge in eine feinkörnige globular-lamellare
$(\alpha+\beta)$-Struktur umzuwandeln, müssen die nachfolgenden
Umformungen unterhalb der Umwandlungstemperatur erfolgen.
Mit zunehmender Annäherung an die Endform wird dabei die
Temperatur nach vorgegebenen Verfahrensplänen erniedrigt,
um eine ausreichende örtliche Umformung in einem Temperaturbereich, der zu der gewünschten Umkristallisation
führt, zu erlangen. Zur Erzielung hoher statischer und
dynamischer Festigkeitswerte ist darauf zu achten, daß
beim Fertigschmieden am ganzen Werkstück ein möglichst
konstanter Umformgrad ($\varphi \geqq 0,3$) bei zugleich niedriger
Temperatur erzielt wird.
Es ist darauf zu achten, daß die in das Werkstück eingebrachte Umformarbeit keine unzulässige Temperaturerhöhung
verursacht. Keinesfalls darf hierdurch die Umwandlungstemperatur überschritten werden [5]. Die Umformgeschwindigkeit ist so zu wählen, daß die Wärmeabgabe vom Werkstück an das Werkzeug etwa durch die beim Umformen entstehende Wärme kompensiert wird.

Bild 8 zeigt Fließkurven der Legierung Ti Al6 Sn2 Zr4 Mo2.

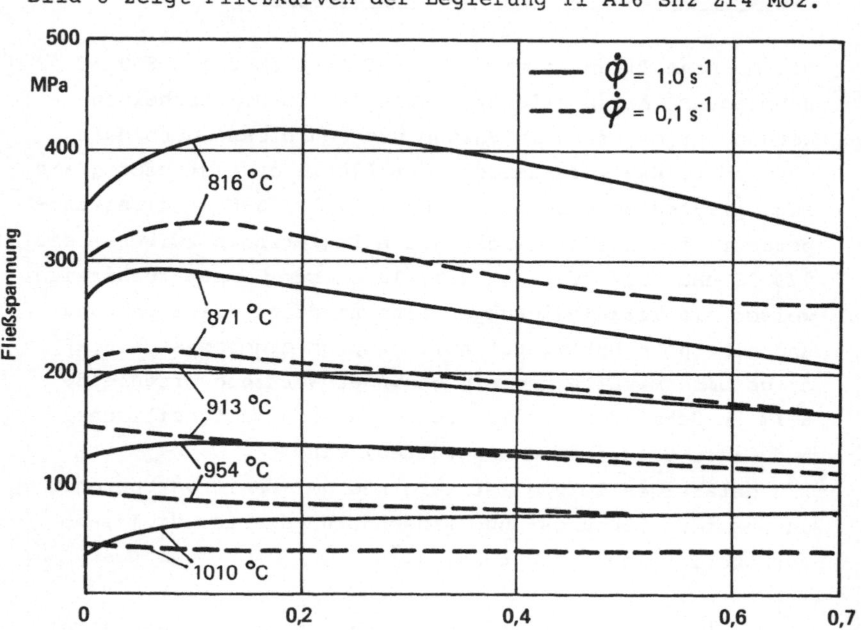

Bild 8: Fließkurven von Ti Al6 Sn2 Zr4 Mo2.

Durch langsames isothermes Umformen bei Temperaturen dicht
unterhalb der Umwandlungstemperatur läßt sich das Fließ-
vermögen von Titanlegierungen gegenüber dem "normalen"
Schmieden stark verbessern. Die Umformung erfolgt dann
unter Ausnützung des Kriechvermögens des Werkstoffs, so
daß sich kleine Wanddicken und scharfe Querschnittsände-
rungen erzielen lassen. Bei hinreichend kleiner Umformgeschwindigkeit und sehr feinkörnigem Gefüge verhalten sich
manche Titanlegierungen superplastisch. Zum isothermen
bzw. superplastischen Umformen sind jedoch (teure) heiz-
bare Gesenke aus gegossenen Superlegierungen erforderlich.

3.4 Umformverhalten von Nickelbasislegierungen

Das Umformverhalten von Ni-Legierungen ist noch schwieriger als das von Ti-Legierungen. Erstere sind speziell für
den Einsatz bei hohen Temperaturen ausgelegt und haben
unter diesen Einsatzbedingungen hohe Festigkeits- und
niedrige Duktilitätswerte. Bei Legierungen mit hohem γ'-
Anteil (z. B. IN 100) muß mit Rücksicht auf die thermische
Belastbarkeit der Werkzeuge häufig unterhalb der γ'-Lö-
sungstemperatur umgeformt werden. Die Schmiedbarkeit ist
dann infolge der hohen Festigkeit und der geringen Duktilität sehr begrenzt. Die Schmiedetemperatur der meisten
Nickellegierungen liegt zwischen ca. 1050 °C und etwa
1200 °C. Langsames Umformen hat den Vorteil niedriger
Kräfte und vergleichsweise hoher Duktilität. Es erfordert
jedoch die Verwendung beheizter, hoch hitzebeständiger
Werkzeuge. Andernfalls würde das Werkstück zu viel Wärme
an das Gesenk abgeben, dieses thermisch überbeanspruchen
und dabei selbst zu rasch abkühlen. Hierbei würde die
Fließspannung steigen und die Duktilität abnehmen.

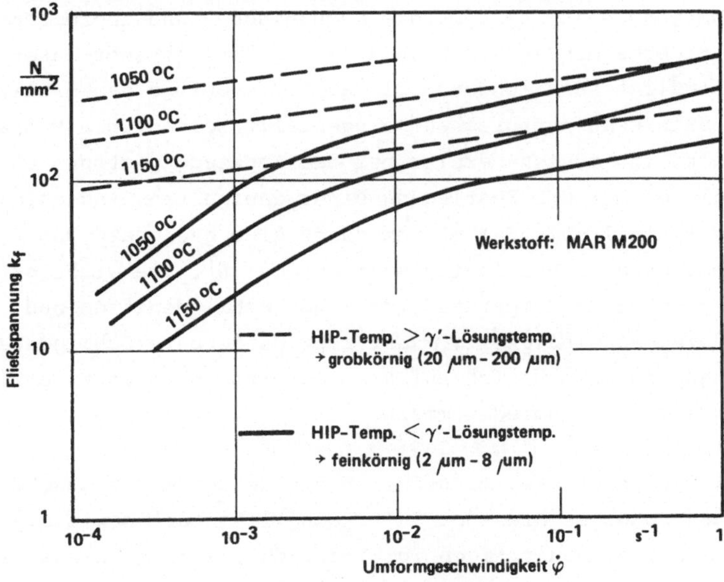

Bild 9: Fließkurven von MAR M 200

Für das isotherme Umformen sind Werkzeuge aus TZ-Molybdän
erforderlich, die nur im Vakuum bzw. unter Schutzgas
betrieben werden dürfen. Dies macht im allgemeinen den
Einsatz von Vakuum-Heißpressen erforderlich.

Bild 9 zeigt Fließkurven der pulvermetallurgisch hergestellten Legierung MAR M 200. Man erkennt, daß die Fließspannung bei feinkörnigem Gefüge niedriger und in stärke-
rem Maße geschwindigkeitsabhängig ist als bei grobem Korn.
Für $\dot{\varphi} \leq 10^{-2}\ s^{-1}$ verhält sich der feinkörnige Werkstoff
superplastisch, da der sog. m-Wert > 0,3 ist.

4. Fertigungsverfahren für Verdichter- und Turbinenscheiben
4.1 Konventionelles Schmieden

Das Schmieden erfolgt meist in mehreren Arbeitsgängen.
Nach dem Anwärmen auf Schmiedetemperatur wird das Rohteil
zunächst vorgestaucht, meist auf einer hydraulischen Presse. Das Fertigschmieden der vorgestauchten Form geschieht
je nach Werkstoff und Scheibengeometrie in einer oder zwei
Schmiedeoperationen. Da die mechanischen und thermischen
Betriebsbeanspruchungen an Scheiben örtlich sehr unterschiedlich sein können, soll auch das Gefüge den örtlichen
Bedürfnissen entsprechend eingestellt werden. Die Maßnah-
men hierzu müssen bereits bei der Werkzeugkonstruktion
beginnen mit der Festlegung einer geeigneten Stadienfolge
und Rohteilform. Sie müssen durch eine geeignete Wahl der
Schmiedeparameter fortgesetzt werden [6]. Die wichtigsten
hiervon sind: Temperatur, Zahl der Schmiedehitzen und
Umformgeschwindigkeit. Exakt reproduzierbare Schmiedebe-
dingungen lassen sich auf Schmiedepressen leichter erzielen als mit Schmiedehämmern.

Bild 10 zeigt etwa maßstäblich die Kontur einer geschmiedeten Scheibe, die sog. "Ultraschall-Kontur" sowie die
Form des einbaufertigen Turbinenrades. Infolge des schlech-
ten Umformvermögens der Superlegierungen und der starken

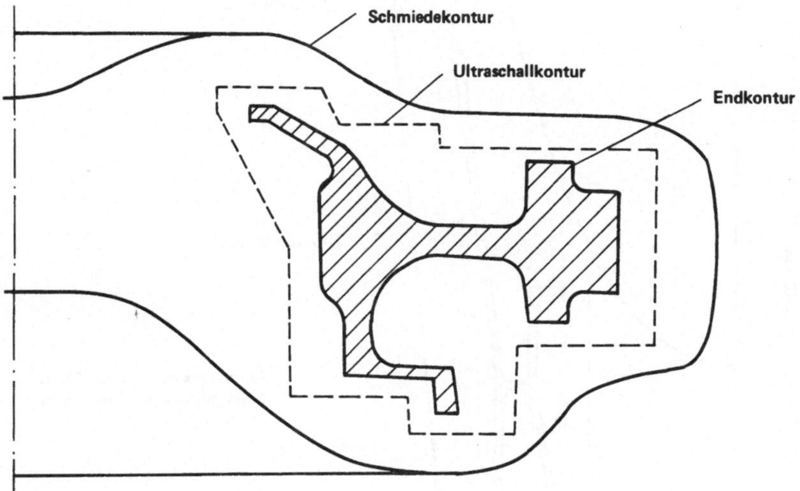

Bild 10: Turbinenscheibe aus Waspaloy

Abkühlung der Randzone läßt sich beim konventionellen
Schmieden mit nur mäßig vorgewärmten Gesenken keine bes-
sere Annäherung der Schmiedekontur an die Fertigkontur des
Teiles erreichen. Im vorliegenden Beispiel, das noch
keinesfalls besonders ungünstig ist, werden über 80 % des
Einsatzgewichtes zerspant.

Bild 11 zeigt eine Rotorwelle bei der ca. 90 % des Einsatzgewichtes zerspant werden. Das Teil wird zunächst
unter einer hydraulischen Presse vorgestaucht, danach
zwischenerhitzt und auf einem Gegenschlaghammer fertig-
geschmiedet. Ein besonderes Problem bei diesem Bauteil
liegt darin, daß es im nahezu fertigbearbeiteten Zustand
an den Stellen S_1 und S_2 mit anderen Teilen durch Elek-
tronenstrahlschweißen verbunden werden muß. Zum mikro-
rißfreien Schweißen von Waspaloy ist aber ein feinkörniges
Schmiedegefüge erforderlich, das sich im vorliegenden Fall
nur mit erheblichem Aufwand erzielen läßt.

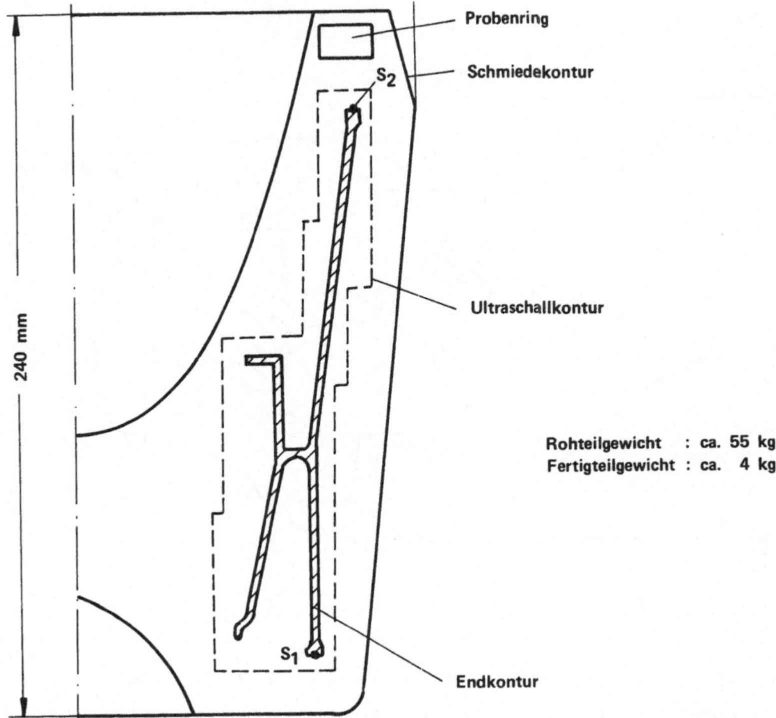

Bild 11: Konventionell geschmiedeter Rotor aus Waspaloy.

4.2 "Moderne" Verfahren zur Scheibenherstellung

Die vorangegangenen Beispiele haben gezeigt, daß sich
unter Verwendung schmelzmetallurgisch hergestellten Aus-
gangsmaterials und konventioneller Schmiedetechnik nur
Schmiede-Endformen erzeugen lassen, die gegenüber dem
einbaufertigen Teil ein erhebliches Aufmaß besitzen.
Angesichts steigender Werkstoffpreise und dem zunehmenden
Zwang zur rohstoffsparenden, wirtschaftlichen Fertigung
gewinnen die Verfahren des "Heiß"-Umformens bzw. das
isotherme Umformen, zunehmend an Bedeutung. Bei diesen
Verfahren werden die Gesenke auf so hohe Temperaturen
vorgewärmt, daß nur noch ein geringer Temperaturgradient
zwischen Gesenk und Werkstück (beim Heißschmieden) bzw.
gar keiner (beim isothermen Umformen) vorliegt. Hierzu
sind teure Gesenke entweder aus gegossenen Superlegierungen

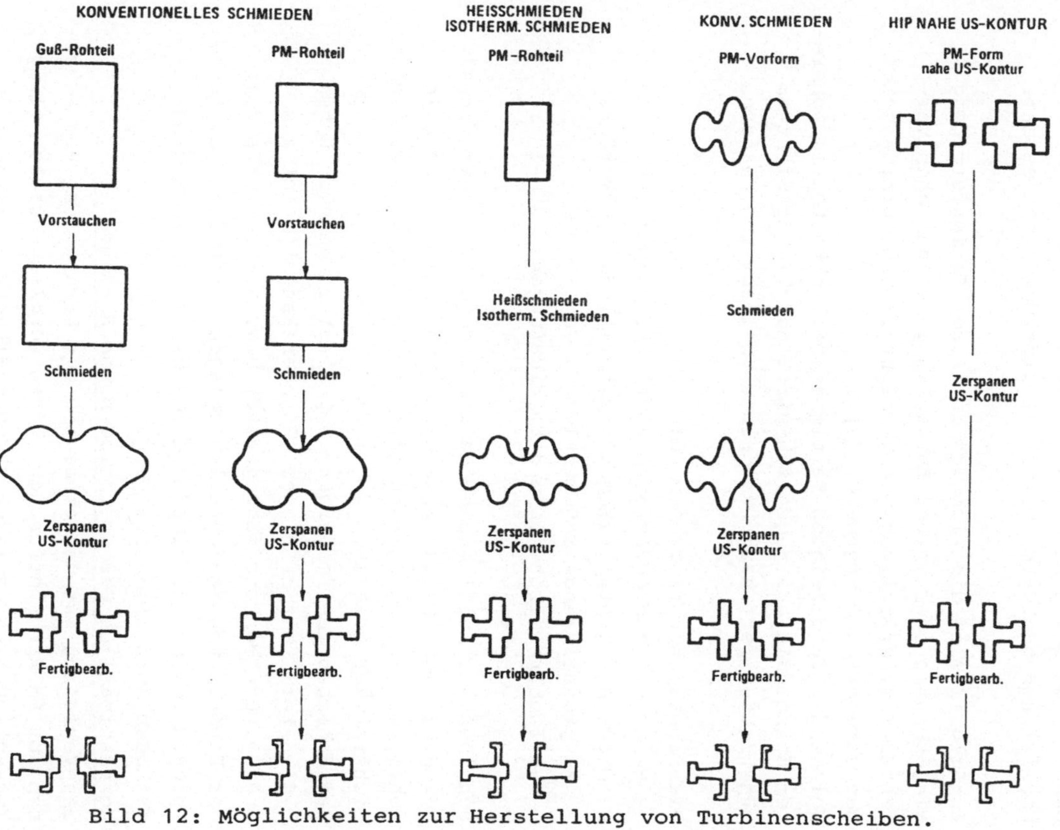

Bild 12: Möglichkeiten zur Herstellung von Turbinenscheiben.

oder aus TZ-Molybdän nötig. Letztere erfordern ein Arbeiten unter sehr reinem Schutzgas oder im Vakuum. Die genannten Verfahren sind insbesondere für die im Vordringen
begriffenen hochwarmfesten PM-Legierungen interessant.

In Bild 12 sind die wichtigsten in der Serienfertigung
eingesetzten oder in der Entwicklung befindlichen Verfahren zur Herstellung von Verdichter- und Turbinenscheiben schematisch aufgezeigt [7].
Welche der aufgezeigten Fertigungsmöglichkeiten im konkreten Einzelfall aus technisch-wirtschaftlicher Sicht am
günstigsten ist, hängt von einer Vielzahl von Faktoren ab.

Die wichtigsten hiervon sind:
- Werkstoff und geforderte Bauteileigenschaften
- Größe und Gestalt (Kompliziertheit) des Werkstücks
- Zu produzierende Stückzahl
- Verfügbare Maschinen und Einrichtungen.

Scheiben aus Knetlegierungen wie Inconel 718 oder Waspaloy
werden in den meisten Fällen noch durch konventionelles
Schmieden von vakuumerschmolzenen Gußblöcken am wirtschaft-
lichsten hergestellt.
Bei Scheiben aus "modernen" Superlegierungen mit hohem
γ'-Anteil, wie z. B. Merl 76 oder René 95, ist aus
Gründen der Werkstoffhomogenität bzw. der Umformbarkeit
der Einsatz von PM-Rohteilen zwingend erforderlich. Nach-
folgend sollen die sich hierfür anbietenden Fertigungs-
methoden kurz erläutert werden:
<u>Konventionelles Gesenkschmieden von PM-Rohteilen</u>
Das Verfahren ist das gängigste zur Herstellung von Scheiben aus PM-Werkstoffen. Es können vergleichsweise billige,
konventionelle Schmiedewerkzeuge eingesetzt werden. Besonders wirtschaftlich ist diese Methode bei mittleren
Stückzahlen (einige 10^2) und nicht zu großen Bauteilen
$(d_{max} \approx 400$ mm$)$.

Heißschmieden, isothermes bzw. superplastisches Schmieden

Bei diesem Verfahren werden auf hohe Temperaturen vorge-
heizte teure Werkzeuge eingesetzt. Die Temperaturdifferenz
zwischen Werkstück und Werkzeug liegt i. a. zwischen
200 °C und 300 °C. Hieraus resultiert gegenüber dem konventionellen Schmieden ("kaltes" Gesenk, große Umformgeschwindigkeit) eine starke Verringerung der benötigten
Preßkraft bei gleichzeitig verbesserter Duktilität des
Werkstoffes. Dadurch ist die Möglichkeit gegeben, das
Rohteilgewicht zu vermindern und die Schmiede-Endkontur
besser an die Kontur des Fertigteils anzupassen.

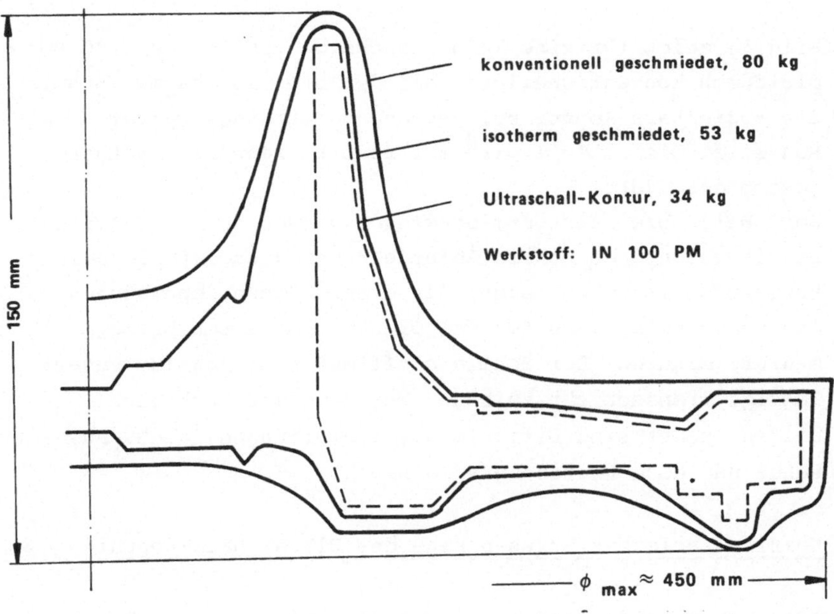

Bild 13: Vergleich der durch konventionelles bzw.
isothermes Schmieden erzielbaren Form einer Turbinenscheibe (nach Pratt & Whitney).

Beim isothermen Umformen werden die Gesenke bis auf Um-
formtemperatur (1050 °C bis 1200 °C) vorgewärmt. Der
Werkstoff verhält sich infolgedessen duktiler als beim
Heißschmieden, so daß die Schmiedeendform noch mehr an die
Form des fertig bearbeiteten Werkstückes angenähert werden
kann. Bei sehr feinkörnigem Gefüge (Korn-Ø wenige µm)
verhalten sich verschiedene Superlegierungen superplastisch. In diesem Zustand lassen sich sehr komplizierte
Formen herstellen. Die hohen Werkzeugkosten machen das
Verfahren nur bei großen Stückzahlen ($> 10^3$) und nicht zu
kleinen Werkstücken mit komplizierter Geometrie wirtschaftlich. Dann kompensiert die Ersparnis an Rohteilgewicht bzw. Zerspanungsaufwand die hohen Werkzeugkosten.

Bild 13 zeigt für eine Turbinenscheibe aus Inconel 100 PM
die durch konventionelles Schmieden bzw. isothermes Schmieden erzielbare Kontur bei Verwendung stranggepreßter
Rohteile. Dieses Teil wird zur Zeit serienmäßig isotherm
geschmiedet [8].
Beim Heiß- bzw. Isotherm-Schmieden werden an die Schmier-
stoffe wesentlich höhere Anforderungen gestellt als beim
konventionellen Schmieden: Die Grenzflächentemperaturen
liegen bei max. 1200 °C, die Druckberührzeiten betragen
mehrere Minuten. Der Schmierstoff muß eine absolut zuverlässige Trennschicht zwischen Werkzeug und Werkstück
bilden, sonst sind Diffusionsverschweißungen, d. h. Aus-
schuß und hohe Kosten unvermeidlich.

Heißisostatisches Pressen nahe der Ultraschall-Kontur

Aus Kostengründen ist man bestrebt, auf das Nachschmieden
nach dem HIP-Prozeß zu verzichten und statt dessen die
HIP-Form möglichst weitgehend an die Form des fertigen
Bauteils anzunähern. Angesichts einer in den letzten
Jahren ständig verbesserten Pulverqualität und sehr stren-
ger Qualitätssicherungsmaßnahmen (Prozeßkontrolle) hat
dieses Verfahren heute die Serienreife erreicht. Trotzdem

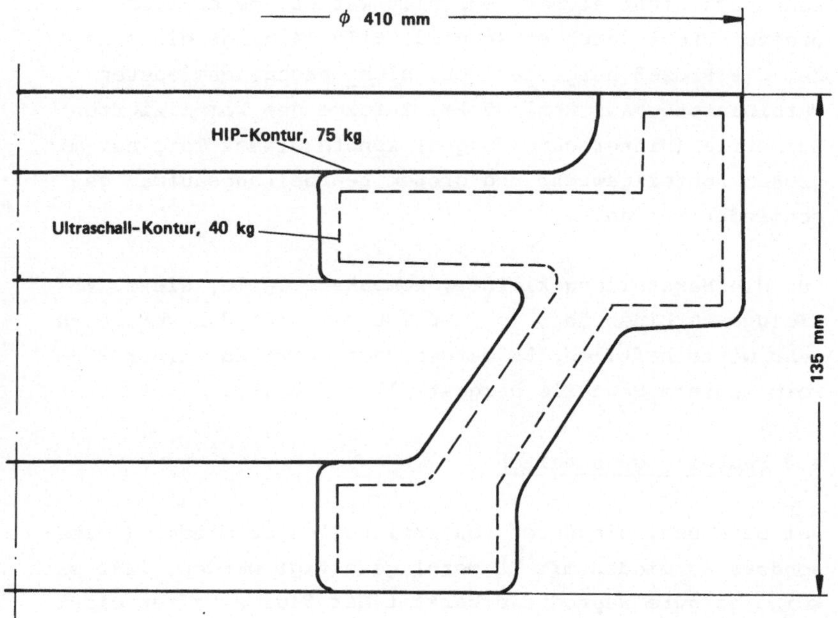

Bild 14: Heißisostatisch gepreßtes Turbinenrad aus
René 95 PM (nach General Electric).

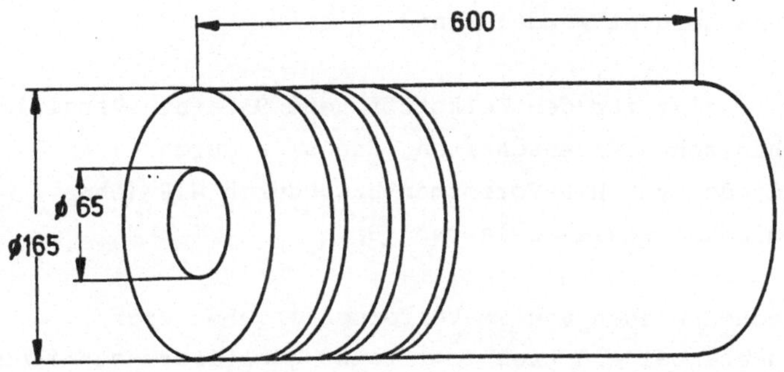

Bild 15: Herstellung von mehreren Kleingasturbinenrädern aus einer HIP-Form (nach General Electric).

kann noch nicht sicher beurteilt werden, ob es sich auf
breiter Front durchsetzen wird. Bild 14 zeigt einen nach
dem HIP-Prozeß hergestellten, <u>nicht</u> nachgeschmiedeten
Turbinenrotor aus René 95 PM. Infolge der komplizierten
Geometrie (Hinterschneidungen) könnte dieses Teil nur mit
großem Rohteilgewicht und großem Bearbeitungsaufmaß geschmiedet werden.

Für die Herstellung kleiner, kompakter Teile, wie z. B.
Kleingasturbinenräder (d_{max} < 250 mm), ist das Verfahren
sehr wirtschaftlich. Es werden dann meist aus einer HIP-
Form mehrere Bauteile hergestellt (Bild 15).

4.3 Bauteileigenschaften

Bei Scheiben, die durch konventionelles Schmieden (insbesondere Schmieden mit Hämmern) gefertigt werden, läßt sich
keine so gute Reproduzierbarkeit der Eigenschaften erreichen wie bei isotherm geschmiedeten Teilen. Bei letzteren
führt die genaue Verfahrenskontrolle mit einer nahezu
konstanten Temperatur innerhalb des Werkstückes zu einer
großen Gleichmäßigkeit der Bauteileigenschaften. Hinzu
kommt, daß beim isothermen Umformen vergleichsweise kon-
turnah geschmiedet werden kann.

Tabelle 3 zeigt für den Werkstoff René 95 einen Vergleich
der mechanischen Eigenschaften, die sich durch
Nachschmieden von HIP-Vorformen bzw. durch HIP (ohne
Nachschmieden) erzielen lassen [8].

Durch Nachschmieden von PM-Verformen ergeben sich
insbesondere bei erhöhter Temperatur günstigere statische
Festigkeitseigenschaften als durch HIP ohne Nachschmieden.
Dies gilt für die bei Turbinenscheiben äußerst wichtigen
LCF-Eigenschaften noch in verstärktem Maße (Vergl. Tabel-
le 4).

statischer Zugversuch

Temp. °C	HIP + Konv. Schmieden				HIP			
	Rp0,2 MPa	Rm MPa	A %	Z %	Rp0,2 MPa	Rm MPa	A %	Z %
20	1256	1658	14,9	16,1	1242	1653	19,7	21,1
650	1182	1517	14,0	17,0	1132	1500	13,0	14,4

Zeitstandversuch, 650 °C, σ = 1034 MPa

HIP + Konventionelles Schmieden		HIP	
Zeit bis Bruch h	Dehnung %	Zeit bis Bruch h	Dehnung %
146,5	11,6	46,0	5,4
40,3	7,4	62,8	4,7

Tabelle 3: René 95PM. Festigkeit von Kleingasturbinenrädern [8].

statischer Zugversuch

Temp. °C	HIP + isotherm. Schmieden				HIP			
	Rp0,2 MPa	Rm MPa	A %	Z %	Rp0,2 MPa	Rm MPa	A %	Z %
20	1268	1668	20,4	21,1	1249	1661	18,5	20,1
650	1143	1494	16,1	15,6	1126	1496	12,3	14,5

Zeitstandversuch, 650 °C, σ = 1034 MPa

HIP + isothermes Schmieden		HIP	
Zeit bis Bruch h	Dehnung %	Zeit bis Bruch h	Dehnung %
47,5 *	4,6	29,0	5,0

* Mittelwert aus 4 Messungen

LCF, dehnungsgesteuert, 433 °C

Dehnung	HIP + isoth. Schmied.		HIP			
	LW bis Bruch		LW bis Bruch			
0,78 %	$66,7 \cdot 10^3$	$58,6 \cdot 10^3$	$18,2 \cdot 10^3$	$32,2 \cdot 10^3$	$28,4 \cdot 10^3$	$28,7 \cdot 10^3$
0,66 %	$112,4 \cdot 10^3$	$225,7 \cdot 10^3$	$68,2 \cdot 10^3$	$120,6 \cdot 10^3$	—	—

Tabelle 4: René 95 PM. Eigenschaften bei HIP bzw. HIP plus isothermem Schmieden [8].

4.4 Entwicklungstendenzen

Zukünftig werden noch erhebliche Anstrengungen zur Kosten-
ersparnis bei der Scheibenherstellung erforderlich sein,
vor allem auf dem Wege der Verminderung des Einsatzgewichtes.
Zur Zeit werden aus Gründen der US-Prüfbarkeit nur gerad-
linige Ultraschallkonturen verwendet. Das allseitige
Aufmaß gegenüber der fertig bearbeiteten Endkontur beträgt
2,5 - 3,0 mm. In den nächsten Jahren wird dieses Aufmaß
infolge verbesserter Prüfmethoden auf ca. 1,5 mm vermindert werden können. Es werden dann auch gekrümmte Mantel-
linien als US-Kontur möglich sein.

Zur Zeit lassen sich durch isothermisches Schmieden in der
Serienfertigung Konturen erzeugen, die gegenüber der
US-Kontur ca. 1,5 mm Aufmaß haben. Hier sind für die Zukunft kaum weitere Werkstoffeinsparungen zu erwarten.
HIP-Konturen hingegen haben unter Produktionsbedingungen
zur Zeit ein Aufmaß von 4 - 5 mm gegenüber der US-Kontur;
im Versuchsstadium sind HIP-Formen mit 1,5 - 2,5 mm.

Bei den Pulvern geht der Trend zu noch größerer Reinheit
und kleineren Partikeldurchmessern. Sie liegen heute zum
Teil schon unterhalb von 50 µm.

Schrifttum

1   Dander, W.: Das Ermüdungsrißwachstum in bruchmechani-
     scher Darstellung. Z. Werkstofftechn. 13 (1982),
     S 385 - 388.
2   Eßlinger, P.: Bewertungskriterien für Werkstoffe des
     Flugtriebwerkbaues. Metall 36 (1982) 6, S. 654 - 659.
3   van Kann, H.: Der Einfluß der Verformungsbedingungen
     beim Gesenkschmieden auf die mechanischen Eigenschaf-
     ten von Gesenkschmiedestücken aus Titan-Legierungen.
     TZ für praktische Metallbearbeitung, 1967.

4   Kramer, K.-H.: Herstelltechnologien von Titan und
     Titanlegierungen. Metall 36 (1982) 6, S. 659 - 668.

5   Semiatin, S.L. Lahoti, G.D.: Deformation und unstab-
     le flow in hot forging of Ti6 Al2Sn 4Zr 2Mo 0.1Si.
     Metallurg. Trans., Vol 12A (1981), S. 1705 - 1717.

6   Schröder, G., Boer, C.R.: Optimierung des Schmiedens
     von Nickelbasislegierungen. wt-Z. ind. Fertig. 72
     (1982), S. 575 - 578.

7   Hansen, W., Wilhelm, H.: Hersteller und Verarbeiter
     im Spannungsfeld der Zusammenarbeit.
     in: Neue Verfahren der Massivumformung. Vortrags-
     texte eines Symposiums der Deutschen Gesellschaft
     für Metallkunde, Oktober 1981. ISBN3-88355-061-2.

8   Coyne, E.: Superalloy Turbine Components - which is
     the Superior Manufacturing Process: as - HIP, HIP +
     Isoforge, or Gatorizing of Extrusion consolidated
     Billet. Vortrag, Metal Powder Report Conference,
     "Aerospace Materials for the 1980's", Zürich, Novem-
     ber 1980.

Berichte aus dem Institut für Umformtechnik
der Universität Stuttgart

Herausgeber Professor Dr.-Ing. Kurt Lange

1   Untersuchung über den Einfluß der Belastungszeit auf die Streuung der Rückfederung von Biegeteilen
Von Dipl.-Ing. Klaus Tafel. 70 Seiten Text u. 64 Seiten mit 49 Bildern u. 15 Tafeln.   Vergriffen

2/3   Untersuchungen über das freie Napfen
Von Dipl.-Ing. Gerhard Schmitt und Dipl.-Ing. Dieter Schmoeckel.
Untersuchungen über den Kraft- und Arbeitsbedarf sowie den Umformwirkungsgrad beim Vorwärts-Vollfließpressen von Stahl
Von Dipl.-Ing. Dieter Kast. 40 Seiten Text u. 43 Seiten mit 47 Bildern u. 5 Tafeln.   28,— DM

4   Untersuchungen über die Werkzeuggestaltung beim Vorwärts-Hohlfließpressen von Stahl und Nichteisenmetallen
Von Dipl.-Ing. Dieter Schmoeckel. 72 Seiten Text u. 117 Seiten mit 179 Bildern.   39,— DM

5   Untersuchungen über das Stauchen und Zapfenpressen
Von Dipl.-Ing. Märten Burgdorf. 126 Seiten Text u. 58 Seiten mit 138 Bildern u. 4 Tafeln.   55,— DM

6   Untersuchungen über die Streuung der Kräfte und Arbeiten beim Fließpressen in der laufenden Fertigung und den Einfluß der Phosphatschichtdicke und des Schmiermittels
Von Dipl.-Ing. Hans-Dietrich Witte. 38 Seiten Text u. 48 Seiten mit 49 Bildern.   30,— DM

7   Untersuchungen über das Rückwärts-Napffließpressen von Stahl bei Raumtemperatur
Von Dipl.-Ing. Gerhard Schmitt. 132 Seiten Text u. 93 Seiten mit 130 Bildern u. 5 Tafeln.   34,— DM

8   Die Abbildegenauigkeit beim Biegen im 90°-V-Gesenk und ihre Beeinflussung durch Nachdrücken im Gesenk durch Nachdrücken im Gesenk
Von Dipl.-Ing. Eckart Dannenmann. 50 Seiten Text u. 31 Seiten mit 28 Bildern u. 1 Tafel.   Vergriffen

9   Untersuchungen über den Zusammenhang zwischen Vickershärte und Vergleichsformänderung bei Kaltumformvorgängen
Von Dipl.-Ing. Hans Wilhelm. 50 Seiten Text u. 35 Seiten mit 37 Bildern u. 2 Tafeln.   Vergriffen

10   Untersuchungen über das Abstreckziehen von zylindrischen Hohlkörpern bei Raumtemperatur
Von Dipl.-Ing. Rolf K. Busch. 86 Seiten Text u. 92 Seiten mit 97 Bildern.   Vergriffen

11   Vorgänge beim elektromagnetischen und elektrohydraulischen Umformen von metallischen Werkstücken
Von Dipl.-Ing. Herbert Müller. 90 Seiten Text u. 110 Seiten mit 93 Bildern u. 10 Tafeln.   22,— DM

12   Ein Verfahren zur näherungsweisen Berechnung des Spannungs- und Formänderungszustandes beim Fließen starrplastischer Werkstoffe
Von Dipl.-Ing. Gerhard Adler. 124 Seiten Text u. 76 Seiten mit 72 Bildern.   Vergriffen

13   Modellgesetzmäßigkeiten beim Rückwärtsfließpressen geometrisch ähnlicher Näpfe
Von Dipl.-Ing. Dieter Kast. 101 Seiten Text u. 73 Seiten mit 60 Bildern u. 6 Tafeln.   Vergriffen

14   Untersuchungen über das Genauschneiden von Stahl und Nichteisenmetallen
Von Dipl.-Ing. Wilfried Krämer. 96 Seiten Text u. 132 Seiten mit 128 Bildern u. 10 Tafeln.   Vergriffen

15   Entwicklung und Erprobung eines Simulators zur reproduzierbaren Nachahmung der Kraft-Weg-Verläufe von Umformvorgängen
Von Dipl.-Ing. Kurt Schmid. 88 Seiten Text u. 38 Seiten mit 35 Bildern u. 2 Tafeln.   17,— DM

16   Walzrichten von Metallbändern mit symmetrisch angestellter Fünf-Walzen-Richtmaschine
Von Dipl.-Ing. Hans-Dietrich Witte. 108 Seiten Text u. 63 Seiten mit 60 Bildern u. 8 Tafeln.   22,— DM

17/18   Erzeugung räumlicher Blechgebilde mittels Flächenbiegung
Konstruktion, Abwicklung und Herstellung von Schraubtorsen aus Blech
Von Prof. Dr.-Ing. E. h. Dr. techn. h. c. Otto Kienzle.
120 Seiten Text u. 55 Seiten mit 86 Bildern u. 3 Tafeln.   22,— DM

19   Einfluß der Alterung auf die mechanischen Eigenschaften von Stählen zum Kaltfließpressen
Von Dipl.-Ing. Vladimir Hasek, CSc. 43 Seiten Text u. 54 Seiten mit 50 Bildern u. 3 Tafeln.   16,— DM

20   Beitrag zur Frage der Spannungen, Formänderungen und Temperaturen beim axialsymmetrischen Strangpressen
Von Dipl.-Ing. Rolf Dalheimer. 118 Seiten Text u. 76 Seiten mit 79 Bildern u. 3 Tafeln.   Vergriffen

21   Über den Einfluß der Werkzeuggeschwindigkeit auf den Stauchvorgang
Von Dipl.-Ing. H.-J. Metzler. 127 Seiten Text u. 100 Seiten mit 94 Bildern u. 6 Tafeln.   25,— DM

22   Numerische Behandlung von Verfahren der Umformtechnik
Von Dr.-Ing. Elmar Steck. 67 Seiten Text u. 22 Seiten mit 43 Bildern.   16,— DM

23   Ein Verfahren zur näherungsweisen Berechnung der Wärmeentwicklung und der Temperaturverteilung beim Kaltstauchen von Metallen
Von Dipl.-Ing. Walther Pohl. 78 Seiten Text u. 51 Seiten mit 61 Bildern u. 4 Tafeln.   21,— DM

24   Untersuchungen über das Drückwalzen zylindrischer Hohlkörper und Beitrag zur Berechnung der gedrückten Fläche und der Kräfte
Von Dipl.-Ing. Hans-Jürgen Dreikandt. 161 Seiten Text u. 79 Seiten mit 73 Bildern u. 6 Tafeln.   Vergriffen

25   Über den Formänderungs- und Spannungszustand beim Ziehen von großen unregelmäßigen Blechteilen
Von Dipl.-Ing. Vladimir Hasek, CSc. 129 Seiten Text u. 106 Seiten mit 109 Bildern u. 9 Tafeln.   35,— DM

26   Über die Anisotropie des plastischen Verhaltens stranggepreßter Stäbe aus hexagonalen Metallen
Von Dipl.-Ing. Günther Schröder. 129 Seiten Text u. 75 Seiten mit 97 Bildern u. 2 Tafeln.   Vergriffen

27   Die Messung der mechanischen Kontaktspannung in der Wirkfuge Werkzeug — Werkstück bei Umformverfahren
Von Dipl.-Ing. Fritz Dohmann. 99 Seiten Text u. 82 Seiten mit 93 Bildern u. 4 Tafeln.   Vergriffen

28   Beitrag zur rechnerunterstützten Auslegung von Pressengestellen
Von Dipl.-Ing. Manfred Geiger. 94 Seiten u. 56 Seiten mit 63 Bildern.   Vergriffen

29 **Untersuchungen über das Aufweittiefziehen**
Von P. S. Raghupathi. M. E. ISBN 3-7736-0780-6.
80 Seiten Text u. 54 Seiten mit 73 Bildern u. 2 Tafeln.
32.- DM

30 **Faltenbildung als Verfahrensgrenze beim Stauchen von Hohlkörpern**
Von Dipl.-Ing. Klaus Dieterle. ISBN 3-7736-0781-4.
55 Seiten Text u. 35 Seiten mit 43 Bildern u. 3 Tafeln.
28.— DM

31 **Beitrag zur Ermittlung von Fließkurven im kontinuierlichen hydraulischen Tiefungsversuch**
Von Dipl.-Ing. Franc Gologranc. ISBN 3-7736-0785-7.
125 Seiten Text u. 58 Seiten mit 95 Bildern u. 6 Tafeln.
Vergriffen

32 **Untersuchungen an Strangpreßmatrizen**
Von Dipl.-Ing. Klaus Gieselberg. ISBN 3-7736-0786-5.
101 Seiten Text u. 56 Seiten mit 69 Bildern.
45.— DM

33 **Beitrag zur Messung der Strangoberflächentemperatur beim Strangpressen**
Von Dipl.-Ing. Karl-Heinz Friedrich. ISBN 3-7736-0787-3.
83 Seiten Text u. 90 Seiten mit 84 Bildern u. 3 Tafeln.
48.— DM

34 **Über das Umformverhalten von Blechen aus Titan und Titanlegierungen**
Von Dipl.-Ing. Hans Wilhelm. ISBN 3-7736-0788-1.
107 Seiten Text u. 69 Seiten mit 76 Bildern u. 13 Tafeln.
48.— DM

35 **Untersuchung der magnetischen Induktion, Stromdichte und Kraftwirkung bei der Magnetumformung**
Von Dipl.-Ing. Volker Schmidt. ISBN 3-7736-0789-X.
60 Seiten Text u. 53 Seiten mit 84 Bildern.
21.— DM

36 **Der Stofffluß beim kombinierten Napffließpressen**
Von Dipl.-Ing. Rolf Geiger. ISBN 3-7736-0790-3.
111 Seiten Text u. 74 Seiten mit 80 Bildern u. 6 Tafeln.
Vergriffen

37 **Beitrag zum Verhalten superplastischer Werkstoffe beim Massivumformen**
Von Dipl.-Ing. Hans Schelosky. ISBN 3-7736-0791-1.
123 Seiten Text u. 61 Seiten mit 60 Bildern u. 4 Tafeln.
48.— DM

38 **Energieumsatz beim elektrohydraulischen Umformen**
Von Dipl.-Ing. Hans-Joachim Weckerle. ISBN 3-7736-0792-X.
103 Seiten Text u. 46 Seiten mit 56 Bildern.
45.— DM

39 **Elastische Wechselwirkungen an Gestell und Hauptgetriebe weggebundener Pressen**
Von Dipl.-Ing. Lutz Schemperg. ISBN 3-7736-0793-8.
91 Seiten Text u. 58 Seiten mit 65 Bildern u. 3 Tafeln.
45.— DM

40 **Über das plastische Verhalten von Sintermetallen bei Raumtemperatur**
Von Dipl.-Ing. Hartmut Höneß. ISBN 3-7736-0794-6.
84 Seiten Text u. 54 Seiten mit 67 Bildern u. 2 Tafeln.
45.— DM

41 **Untersuchungen zum Halbwarmfließpressen von Stahl**
Von Dr.-Ing. Rolf Geiger, Dipl.-Ing. Eckart Dannenmann und Dipl.-Ing. Jean Stefanakis.
ISBN 37736-0795-4. 50 Seiten Text u. 33 Seiten mit 34 Bildern u. 2 Tafeln.
Vergriffen

42 **Änderung der Werkstoffeigenschaften beim Ziehen von zylindrischen Hohlkörpern aus austenitischen und ferritischen nichtrostenden Stählen**
Von Dipl.-Ing. Rolf Zeller. ISBN 3-7736-0796-2.
80 Seiten Text u. 52 Seiten mit 34 Bildern u. 2 Tafeln.
38.— DM

43 **Untersuchungen über das Fließpressen superplastischer Werkstoffe**
Von Dr.-Ing. Hans Schelosky. ISBN 3-7736-0797-0.
36 Seiten Text u. 24 Seiten mit 26 Bildern u. 1 Tafel.
30.— DM

44 **Umformende Bearbeitung in flexiblen Fertigungssystemen**
Von Dipl.-Ing. Hartmut Kaiser. ISBN 3-7736-0798-9.
87 Seiten Text u. 24 Seiten mit 47 Bildern.
36.— DM

45 **Geometrische Eigenschaften tiefgezogener kreiszylindrischer Näpfe**
Von Dipl.-Ing. Dieter Schlosser. ISBN 3-7736-0799-7.
107 Seiten Text u. 64 Seiten mit 60 Bildern u. 9 Tafeln.
48.-- DM

46 **Die Eigenschaften einer AlZnMgCu-Legierung nach ausgewählten Kombinationen von Wärmebehandlung und Kaltumformung**
Von Dipl.-Ing. Karl Hankele. ISBN 3-7736-0880-2.
86 Seiten Text u. 51 Seiten mit 52 Bildern u. 4 Tafeln.
45.— DM

47 **Kaltmassivumformen von Sintermetall**
Von Dipl.-Ing. Hans Dieter Schacher. ISBN 3-7736-0881-0.
84 Seiten Text u. 44 Seiten mit 47 Bildern u. 5 Tafeln.
42.— DM

48 **Rechnerunterstützte Arbeitsplanerstellung und Kostenrechnung beim Kaltmassivumformen von Stahl**
Von Dipl.-Ing. Peter Noack. ISBN 3-7736-0882-9.
216 Seiten Text u. 116 Seiten mit 134 Bildern u. 23 Tafeln.
65.— DM

49 **Beitrag zur beanspruchungsgerechten Auslegung von rotationssymmetrischen Fließpreßmatrizen**
Von Dipl.-Ing. Günther Krämer. ISBN 3-7736-0883-7.
94 Seiten Text u. 53 Seiten mit 56 Bildern.
48.— DM

50 **Erzeugung gratfreier Schnittflächen durch Aufteilen des Schneidvorgangs (Konterschneiden)**
Von Dipl.-Ing. Heinz Liebing. ISBN 3-7736-0884-5.
87 Seiten Text u. 51 Seiten mit 55 Bildern u. 4 Tafeln.
46.— DM

Die Berichte 1 bis 50 sind zu beziehen durch das Institut für Umformtechnik. Holzgartenstr. 17, 7000 Stuttgart 1

51 **Berechnung der elastischen Eigenschaften von Baugruppen im Pressenbau**
Von Dipl.-Ing. Herbert Blum ISBN 3-540-09804-6.
151 Seiten mit 55 Abbildungen. 48,— DM

52 **Untersuchung der Verfahrensgrenzen beim 180°-Biegen von Fein- und Mittelblechen**
Von Dipl.-Phys. Wolfgang Schaub. ISBN 3-540-09881-X.
65 Seiten mit 24 Abbildungen. 38,— DM

53 **Abstreckgleitziehen von nichtrostenden austenitischen Stählen**
Von Dipl.-Ing. Jobst-H. Kerspe. ISBN 3-540-09882-8.
109 Seiten mit 36 Abbildungen. 43,— DM

54 **Fließpressen von Stahl im Temperaturbereich 773 K (500°C) bis 1073 K (800°C)**
Von Dipl.-Ing. Ulrich Diether. ISBN 3-540-09959-X.
165 Seiten mit 80 Abbildungen. 48,— DM

55 **Die numerisch gesteuerte Radial-Umformmaschine und ihr Einsatz im Rahmen einer flexiblen Fertigung**
Von Dipl.-Ing. Peter Metzger. ISBN 3-540-10073-3.
158 Seiten mit 65 Abbildungen. 43,— DM

56 **Möglichkeiten zur Steuerung des Stoffflusses beim Ziehen großer unregelmäßiger Blechteile**
Von Dr.-Ing. Vladimir V. Hasek. ISBN 3-540-10074-1.
193 Seiten mit 96 Abbildungen. 48,— DM

57 **Beitrag zur Arbeitsgenauigkeit des Kaltmassivumformens**
Von Dipl.-Ing. Herbert Leykamm. ISBN 3-540-10363-5.
165 Seiten mit 84 Abbildungen und 5 Tabellen.. 48,— DM

58 **Untersuchungen über das Verjüngen von zylindrischen Vollkörpern**
Von Dipl.-Ing. Helmut Binder. ISBN 3-540-10466-6.
146 Seiten mit 50 Abbildungen und 3 Tabellen. 43,— DM

59 **Umformverhalten legierter Sintereisen**
Von Dipl.-Ing. Manfred Stilz. ISBN 3-540-11051-8.
170 Seiten mit 75 Abbildungen und 5 Tabellen. 48,— DM

60 **Interaktives Programmsystem zur Erstellung von Fertigungsunterlagen für die Kaltmassivumformung**
Von Dipl.-Ing. Michael Rebholz. ISBN 3-540-11052-6.
121 Seiten mit 46 Abbildungen. 43,— DM

61 **Beitrag zum Ziehen von Blechteilen aus Aluminiumlegierungen**
Von Dipl.-Ing. Michael Blaich. ISBN 3-540-11067-4.
141 Seiten mit 64 Abbildungen und 5 Tabellen. 43,— DM

62 **Auslegung von rotationssymmetrischen Fließpreßwerkzeugen im Bereich elastisch-plastischen Werkstoffverhaltens**
Von Dipl.-Ing. Thomas Neitzert. ISBN 3-540-11623-0.
159 Seiten mit 51 Abbildungen. 53,— DM

63 **Fließpressen von Sintermetall im Temperaturbereich zwischen 873 K (600°C) und 1173 K (900°C)**
Von Dipl.-Ing. Wolfgang Schaub. ISBN 3-540-11678-8.
160 Seiten mit 85 Abbildungen und 9 Tabellen. 53,— DM

64 **Rechnerunterstützte Konstruktion von Umformwerkzeugen und die Fertigungsplanung von Werkzeugelementen**
Von Dipl.-Ing. Dieter Steuss. ISBN 3-540-11856-X.
178 Seiten mit 87 Abbildungen und 6 Tabellen. 53,— DM

65 **Möglichkeiten und Grenzen des Kaltgesenkschmiedens als eine fertigungstechnische Alternative für kleine, genaue Formteile**
Von Dipl.-Ing. Khang Hoang-Vu. ISBN 3-540-11876-4.
156 Seiten mit 62 Abbildungen und 5 Tabellen. 53,— DM

66 **Einsatz numerischer Näherungsverfahren bei der Berechnung von Verfahren der Kaltmassivumformung.**
Von Dipl.-Ing. Karl Roll. ISBN 3-540-11910-8.
166 Seiten mit 49 Abbildungen und 2 Tabellen. 53,— DM

67 **Untersuchung über das Verjüngen von dickwandigen, zylindrischen Hohlkörpern**
Von Dipl.-Ing. Knut Haarscheidt. ISBN 3-540-12229-X.
124 Seiten mit 58 Abbildungen und 6 Tabellen. 58,— DM

68 **Rechnerunterstützte Optimierung des Tiefziehens unregelmäßiger Blechteile**
Von Dipl.-Ing. Hans Glöckl. ISBN 3-540-12522-1.
143 Seiten mit 60 Abbildungen. 58,— DM

69 **Hydrostatisches Fließpressen: Verfahrensparameter und Werkstückeigenschaften**
Von Dipl.-Ing. Jobst H. Kerspe. ISBN 3-540-12537-X.
123 Seiten mit 69 Abbildungen und 5 Tabellen. 58,— DM

70 **Untersuchungen zum Halbwarmfließpressen von Automatenstählen**
Von Dipl.-Ing. Eberhard Nehl. ISBN 3-540-12568-X.
145 Seiten mit 104 Abbildungen. 58,— DM

71 **Entwicklung und Anwendung neuer Schmierstoffprüfverfahren für die Kaltmassivumformung**
Von Dipl.-Ing. Thomas Gräbener. ISBN 3-540-12836-0.
140 Seiten mit 65 Abbildungen. 58,— DM

72 **Einfluß der Blechoberfläche beim Ziehen von Blechteilen aus Aluminiumlegierungen**
Von Dipl.-Ing. Erhard Mössle. ISBN 3-540-12837-9.
142 Seiten mit 62 Abbildungen und 6 Tabellen. 58,— DM

73 **Werkzeugverschleiß in der Massivumformung**
Von Dipl.-Ing. Matthias Weiergräber. ISBN 3-540-13033-0.
72 Seiten mit 36 Abbildungen und 2 Tabellen. 58,— DM

Die Berichte 51 und folgende sind zu beziehen durch den Springer-Verlag, Berlin Heidelberg New York Tokyo

74 **Grundlagen der Umformtechnik I · Fundamentals of Metal Forming Technique I**
298 Seiten. ISBN 3-540-13039-X.   58,— DM

Die Berichte 51 und folgende sind zu beziehen durch den Springer-Verlag, Berlin Heidelberg New York Tokyo